An Introduction to
Metabolic and
Cellular Engineering

An Introduction to
Metabolic and
Cellular Engineering

S Cortassa
CONICET, Argentina

M A Aon
CONICET, Argentina

A A Iglesias
CONICET, Argentina

D Lloyd
University of Wales, UK

World Scientific
New Jersey • London • Singapore • Hong Kong

Published by

World Scientific Publishing Co. Pte. Ltd.

P O Box 128, Farrer Road, Singapore 912805

USA office: Suite 1B, 1060 Main Street, River Edge, NJ 07661

UK office: 57 Shelton Street, Covent Garden, London WC2H 9HE

British Library Cataloguing-in-Publication Data
A catalogue record for this book is available from the British Library.

AN INTRODUCTION TO METABOLIC AND CELLULAR ENGINEERING

ISBN 981-02-4835-0
ISBN 981-02-4836-9 (pbk)

Printed in Singapore by Uto-Print

To Juan Ernesto, Nehuen Quimey, Raul and Pedro. To Gladys (in memoriam).
To Francisca, Maris and Juan Carlos. To Miguel (in memoriam).
To Silvia. To Norberto (in memoriam).

Preface

Metabolic and Cellular Engineering, although as yet only at a beginning, promises huge advances in all fields of the life sciences. The main aim of this book is to introduce students and research workers into this exciting new endeavor. To show a complete picture of the subject, we introduce the main techniques available in the field, in order to point out their power, facilitate their mastery, interpret the achievements already published, and challenge our readers with new problems.

Our own research interests have led us to the elaboration of a wider view on the emergent field of Metabolic Engineering. Thus, here we review the field in order to give a state-of-the-art account. However, in doing this we have selected examples, experiments, and puzzles that, in our opinion, accurately reflect the main advances, achievements, and unsolved problems. So a prospective for the field has also emerged. This book also pretends to be useful to those experimentalists and theoreticians who wish to project themselves into a field that offers great challenges, either experimental or theoretical, for massive integration of the available information.

Until 1960s, metabolic regulation was mainly investigated in isolated and cell-free systems. At present, biotechnology mainly deals with intact cells, and we therefore need to understand how enzymatic reactions behave and are regulated inside the cell. From this standpoint, major limitations arise from the lack of understanding of the behavior of metabolic networks. More precisely; on the one hand, geneticists and molecular biologists produce schemes to explain regulation of gene expression, e.g. by DNA-binding proteins, and on the other hand knowledge of the functioning of metabolic pathways is in some cases fairly complete. However, the link between these two aspects is poorly understood.

Metabolic and Cellular Engineering emphasizes the microorganism (e.g. enzyme function, transport, regulation) and its modification to improve cellular activities, through the use of recombinant DNA. Nevertheless, we assume that the level of performance of the recombinant cells thereby obtained must be evaluated within the context of a specific biotransformation. Thus **Metabolic and Cellular Engineering** is bred of a powerful alliance of two disciplines: Genetics-Molecular Biology and Quantitative Biochemistry and Physiology. Both are driven by continuous refinement of basic understanding of metabolism, physiology, cellular biology (growth, division, differentiation), and the development of new mathematical modeling techniques.

We hope that, even if our aim is minimally attained, then those who have read the book will feel stimulated enough to engage in the field to themselves make new contributions.

The main material of the present book as well as its general structure originated, in part, from a series of lectures given by the authors in the framework of an international course for postgraduate students "Principles of Bioprocess and Metabolic Engineering" supported by the binational centre CABBIO (Centro Argentino-Brasileño de Biotecnología) held at the end of the year 1998 in Chascomús, Buenos Aires. We would like to gratefully acknowledge the participation in that course of Dr. Claudio Voget (Universidad Nacional de La Plata, Argentina) and Dr. Juan Carlos Aon (MIT, USA) for their contribution to the subject of mass and energy transfer and fermentation technology, respectively.

The contribution of the following people for enlightening and useful discussions is gratefully acknowledged: Sam Vaseghi (CITAG, Hamburg, Germany), late Manfred Rizzi (University of Stuttgart, Germany), Marta Cascante (Universitat de Barcelona, Spain), Francesc Mas (Universitat de Barcelona, Spain), Luis Acerenza (Facultad de Ciencias, Uruguay), Carlos E. Argaraña (Universidad Nacional de Córdoba, Argentina), Matthias Reuss (University of Stuttgart, Germany), A.H. Stouthamer (Free University Amsterdam, The Netherlands), Carlos Mignone (Universidad de La Plata, Argentina), Daniel Guebel (Universidad Nacional de Quilmes, Argentina), Nestor V. Torres Darias (Universidad de La Laguna, Spain).

We would also like to express our gratitude for the financial support provided by Fundación Antorchas (ARGENTINA), Centro Argentino Brasileño de Biotecnología (CABBIO), Secretaría de Ciencia y Técnica through the grant ANPCyT PICT'99 1-6074, Facultad de Bioquímica y Ciencias Biológicas (Universidad Nacional del Litoral) and Consejo Nacional de Investigaciones Científicas y Técnicas (CONICET) in Argentina and the Cardiff School of Biosciences, Cardiff University in Wales, United Kingdom.

Chascomús, Buenos Aires
Cardiff, Wales
November, 2001

Contents

Preface   vii

List of Abbreviations   xv

Introduction   1
    Introductory Outlines   1
    Metabolic and Cellular Engineering in the Context of Bioprocess
    Engineering   2
    Tools for Metabolic and Cellular Engineering   3
    Engineering Cells for Specific Biotransformations   5
    Metabolic Areas that Have Been Subjected to MCE   8
    From DNA Sequence to Biological Function   17
    Temporal and Spatial Scaling in Cellular Processes   21
    Scaling in Microbial and Biochemical Systems   22
    Views of the Cell   24
        Black and Grey Boxes: Levels of Description of Metabolic
        Behavior in Microorganisms   24
    Transduction and Intracellular Signalling   29
    Self-organized Emergent Phenomena   30
    Homeodynamics and Coherence   34

Matter and Energy Balances   39
    Mass Balance   39
        General Formulation of Mass Balance   40
        Integral and Differential Mass Balances   41
        Growth Stoichiometry and Product Formation 42
        Biomass and Product Yields   46
        Electron Balance   47
        Theoretical Oxygen Demand   48
        Opening the "Black Box". Mass Balance as the Basis of
        Metabolic Flux Analysis   55
    Energy Balance   63
        Forms of Energy and Enthalpy   64

Calorimetric Studies of Energy Metabolism   67
Heat of Combustion   68
An Energetic View of Microbial Metabolism   73

Cell Growth and Metabolite Production. Basic Concepts   77
Microbial Growth under Steady and Balanced Conditions   77
Microbial Energetics under Steady State Conditions   84
Growth Kinetics under Steady State Conditions   85
The Dilution Rate   86
The Dilution Rate and Biomass Concentration   86
The Dilution Rate and the Growth-limiting Substrate Concentration 87
Biomass and Growth-limiting Substrate Concentration at the Steady
State   88
Growth as a Balance of Fluxes   91
The Flux Coordination Hypothesis   93
Toward a Rational Design of Cells   96
Redirecting Central Metabolic Pathways under Kinetic or
Thermodynamic Control   97
Thermodynamic or Kinetic Control of Flux under Steady State
Conditions   100
Kinetic and Thermodynamic Limitations in Microbial Systems.
Case Studies   102
Saccharomyces cerevisiae   102
Escherichia coli   105
Increasing Carbon Flow to Aromatic Biosynthesis in *Escherichia
coli*   106

Methods of Quantitation of Cellular "Processes Performance"   111
Stoichiometry of Growth: The Equivalence between Biochemical
Stoichiometries and Physiological Parameters   111
A General Formalism for Metabolic Flux Analysis   114
A Comparison between Different Methods of MFA   115
MFA Applied to Prokaryotic and Lower Eukaryotic Organisms   115
MFA as Applied to Studying the Performance of Mammalian
Cells in Culture   118
Metabolic Fluxes during Balanced and Steady State Growth   119
Bioenergetic and Physiological Studies in Batch and Continuous
Cultures. Genetic or Epigenetic Redirection of Metabolic Flux   120
Introduction of Heterologous Metabolic Pathways   120
Metabolic Engineering of Lactic Acid Bacteria for Optimising
Essential Flavor Compounds Production   123

Metabolic Control Analysis 126
Summation and connectivity theorems 131
Control and Regulation 133
The Control of Metabolites Concentration 134
A Numerical Approach for Control Analysis of Metabolic
Networks and Nonlinear Dynamics 134
The TDA Approach as Applied to the Rational Design of
Microorganisms: Increase of Ethanol Production in Yeast 135
Phase I: Physiological, Metabolic and Bioenergetic Studies of
Different Strains of *S. cerevisiae* 136
Phase II: Metabolic Control Analysis and Metabolic Flux
Analysis of the Strain under the Conditions Defined in Phase I 137
Phases III and IV: To Obtain a Recombinant Yeast Strain with
an Increased Dose of PFK, and to Assay the Engineered Strain
in Chemostat Cultures under the Conditions Specified in Phase I 140
Appendix A 142
A Simplified Mathematical Model to Illustrate the Matrix
Method of MCA 142
Appendix B 144
Conditions for Parameter Optimization and Simulation of the
Mathematical Model of Glycolysis 144

Dynamic Aspects of Bioprocess Behavior 145
Transient and Oscillatory States of Continuous Culture 145
Mathematical Model Building 145
Transfer-Function Analysis and Transient-Response Techniques 151
Theoretical Transient Response and Approach to Steady State 152
Transient Responses of Microbial Cultures to Perturbations of the
Steady State 155
Dilution Rate 155
Feed Substrate Concentration 155
Growth with Two Substrates 156
Temperature 156
Dissolved Oxygen 156
The Meaning of Steady State Performance in Chemostat Culture 157
Oscillatory Phenomena in Continuous Cultures 157
1. Oscillations as a Consequence of Equipment Artifacts 157
2. Oscillations Derived from Feedback Between Cells and
Environmental Parameters 158
3. Oscillations Derived from Intracellular Feedback Regulation 159
4. Oscillations Derived from Interactions between Different
Species in Continuous Culture 165
5. Oscillations Due to Synchronous Growth and Division 165

Bioprocess Development with Plant Cells   171
    MCE in Plants: Realities and Potentialities   172
        Plant Transformation for Studies on Metabolism and Physiology 172
    Improving Plants through Genetic Engineering   173
        Improving Plant Resistance to Chemicals, Pathogens and
        Stresses   173
        Improving Quality and Quantity of Plant Products   176
    Using Plant Genetic Engineering to Produce Heterologous Proteins 179
    Tools for the Manipulation and Transformation of Plants   180
    Plant Metabolism: Matter and Energy Flows and the Prospects of
    MCA   183
    Metabolic Compartmentation in Plant Cells   184
    Carbon Assimilation, Partitioning, and Allocation   186
    Carbon Fixation in Higher Plants   188
    MCA Studies in Plants   194
    Regulation and Control: Starch Synthesis, a Case Study   196
    Concluding Remarks   199

Cellular Engineering   201
    Outline   201
    The Global Functioning of Metabolic Networks   202
        The Nature of the Carbon Source Determines the Activation of
        Whole Blocks of Metabolic Pathways with Global Impact on
        Cellular Energetics   203
        Carbon Sources that Share Most Enzymes Required to
        Transform the Substrates into Key Intermediary Metabolites
        under Similar Growth Rates, Bring About Similar Fluxes
        through the Main Amphibolic Pathways   203
        Interaction between Carbon and Nitrogen Regulatory Pathways
        in *S. cerevisiae*   204
        Flux Redirection toward Catabolic (Fermentation) or Anabolic
        (Carbohydrates) Products May Be Generated as a Result of
        Alteration in Redox and Phosphorylation Potentials   206
        Temperature-Dependent Expression of Certain Mutations
        Depend upon the Carbon Source   207
        There Seems to Exist a General Pattern of Control of the
        Intracellular Concentration of Metabolites   207
        Dependence of the Control of Glycolysis on the Genetic
        Background and the Physiological Status of Yeast in Chemostat
        Cultures   211
    Cellular Engineering   212
        Growth Rate, G1 Phase of the Cell Cycle, Production of

Metabolites and Macromolecules as Targets for Cellular
Engineering   213
Catabolite Repression and Cell Cycle Regulation in Yeast   215
Protein Production as a Function of Growth Rate   217
The Selective Functioning of Whole Metabolic Pathways Is
Permissive for Differentiation   220

Bibliography   223

Index   243

List of Abbreviations

AcCoA	acetyl Coenzyme A
ADPGlc	ADP glucose
ADPGlcPPase	ADP glucose pyrophosphorylase
αKG	α ketoglutarate
ADH	alcohol dehydrogenase
ATPase	ATP hydrolase
BST	Biochemical system theory
CDC	cell division cycle
CER	CO2 evolution rate
CSTR	continuously stirred tank reactor
CAM	crassulacean acid metabolism
2DG	2-deoxyglucose 2DG
DAHP	3-deoxy-D-arabino-heptulosonate-7-phosphate
DHAP	dihydroxyacetone-P
D	dilution rate
DBA	Dynamic bifurcation analysis
E4P	erythrose 4 phosphate
EPSP	5-enolpyruvylshikimate-3-phosphate
C_{Ek}^{Ji}	Flux control coefficient
FCH	Flux coordination hypothesis
FBPase	fructose-1,6-bisP phosphatase
G6P	glucose 6 phosphate
GAP	glyceraldehyde-3P
GC/MS	gas chromatography/mass spectrometry
GAPDH	glyceraldehyde 3 phosphate dehydrogenase
HK	hexokinase
K_m	Michaelis Menten constant in enzyme kinetics
MCE	Metabolic and cellular engineering
MCA	Metabolic control analysis
MFA	Metabolic flux analysis
C_{Ek}^{Mi}	Metabolite concentration control coefficient
μ	growth rate
MTP	microtubular protein

m_S	maintenance coefficient
NMR	nuclear magnetic resonance
NADP GDH	NADP dependent glutamate dehydrogenase
OAA	oxalacetate
ODE	ordinary differential equations
OUR	oxygen uptake rate
PP pathway	pentose phosphate pathway
PEP	phosphoenolpyruvate
PEPCK	phosphoenolpyruvate carboxy kinase
PEPCase	phosphoenolpyruvate carboxylase
PFK	phosphofructokinase
PGI	phosphoglucoisomerase
3PG or 3PGA	3 phosphoglycerate
PGK	phosphoglycerokinase
PGM	phosphoglyceromutase
PTS	phosphotransferase system
PEG	polyethylene glycol
PY	pyruvate
qCO2	specific rate of carbon dioxide production
qEtOH	specific rate of ethanol production
qGlc	specific rate of glucose consumption
qO2	specific rate of oxygen consumption
R5P	ribose 5 phosphate
RPPP	reductive pentose phosphate pathway
RQ	respiratory quotient
C_p	specific heat
TP	triose phosphate
TDA	Transdisciplinary approach
TCA cycle	tricarboxylic acid cycle
Vmax	maximal rate in enzyme kinetics
Y_{ATP}	yield of biomass on ATP
Y_{O2}	yield of biomass on oxygen
Y_{PC}	yield of product on carbon substrate
Y_{XC} or Y_{XS}	yield of biomass on carbon substrate
γ_B	degree of reduction of biomass
γ_S	degree of reduction of substrate

Chapter 1

Introduction

Introductory Outlines

In the present book we aim to develop ideas that allow us to confront and solve many problems in Metabolic and Cellular Engineering (MCE). This is not yet a mature speciality, as it is a new area at a meeting point of several disciplines. Evidence for this is provided by the fact that many of the qualitative and quantitative methodologies used are still under development, although in the past ten years striking progress has been achieved (Stephanopoulos *et al.*, 1998; Lee and Papoutsakis, 1999). Nevertheless, there is not yet a well established link between the different disciplines and techniques that are employed in these problems; major inputs come from Molecular Biology, Fermentation Technology and Mathematical Modeling.

We intend to go beyond the assertion that "At present, metabolic engineering is more a collection of examples than a codified science" (Bailey, 1991). However, a valuable and rich experience has accumulated following the explosive development of MCE in the last years. This experience not only reflects the specific achievements (e.g. production of metabolites and heterologous proteins, introduction of heterologous metabolic pathways into microorganisms to give them the ability to degrade xenobiotics, modification of enzymatic activities for metabolite production) but has also allowed us to understand previously unknown aspects of cell function and the regulation of networks of chemical reactions inside cells. The latter places MCE at the interphase between basic and technological research through an iterative, self-correcting and self-fed process for solving the multiple challenges posed by the present developments in biotechnology.

Metabolic and Cellular Engineering in the Context of Bioprocess Engineering

Traditionally, bioprocesses are the bases of the food and pharmaceutical industries. The operation of bioprocesses deals with microbial, plant or mammalian cells, or their components such as enzymes, which are used in the manufacturing of new products, or for the degradation of toxic wastes. The use of microorganisms for the production of fermented foods has a very long history. Since early times, many different bioprocesses have been developed to give an enormous variety of commercial products from cheap ones (e.g. ethanol or organic solvents) to expensive ones (e.g. antibiotics, therapeutic proteins or vaccines). Enzymes and microorganisms such as bakers' yeast are also commercial products which are obtained through different bioprocesses.

Recently, the bases of the broad and highly multidisciplinary field of Metabolic Engineering (ME) have been established (see Bailey, 1991; Cameron and Tong, 1993; Farmer and Liao, 1996; Cameron and Chaplen, 1997; for reviews). ME within the context of Bioprocess Engineering constitutes a thorough transdisciplinary effort toward the development of rationally-designed cells with specific biotransformation capabilities. This transdisciplinary effort requires the participation of scientists with different strengths or abilities in their scientific backgrounds.

The research field of ME was highlighted as an exciting new endeavor in biotechnology in the 1998 International Conference. As a new approach for rationally designing biological systems it is becoming ever more important for biotechnological production processes and medicine. ME can be defined as the introduction of specific modifications to metabolic networks for the purpose of improving cellular properties. Because the challenge of this interdisciplinary effort is to redesign complex biosystems, a rigorous understanding of the interactions between metabolic and regulatory networks is critical. At this point we will adopt the notation Metabolic and Cellular Engineering (MCE) all throughout the book, since it describes more accurately the panoply of activities being undertaken in the field (see below). As such, an important component of MCE is the emphasis of the regulation of metabolic reactions in their cellular entirety (i.e. in the whole organism). This goal differentiates the field from those related areas of life science that adopt the reductionistic approach. Concepts and methodologies of MCE have potential value in the direct application of metabolic design of cellular systems for biotechnology production processes. These include cell-based processes as well as gene therapies, and degradation of recalcitrant pollutants. They promise a great impact on many areas of medicine (Yarmush and Berthiaume, 1997).

At present the following subjects are actively researched in MCE (Cameron and Tong, 1993; Lee and Papoutsakis, 1999) (see below):

1. Experimental and Computational Tools.
2. Applications to the Production of Pharmaceuticals.
3. Applications to Fuels and Chemicals.
4. Biomaterials.
5. Applications to Plants.
6. Applications to Production of Proteins.
7. Evolutionary Strategies for Strain Improvement via ME
8. Medical Applications and Gene Therapy.
9. Higher Level Metabolic Engineering through Regulatory Genes.
10. Environmental Applications.

Tools for Metabolic and Cellular Engineering

MCE requires the development of several tools in the various disciplines contributing to the field. The area of molecular biology needs:

1. Transformation systems for microorganisms used in industrial production, or in bioprocesses (e.g. for *Corynebacterium*, commonly used for the production of aminoacids, or for Pseudomonads, currently used in the degradation of xenobiotics) (see Keasling, 1999, for a review of gene expression tools in bacteria).

2. Promoters and special vectors used in such transformations: e.g. the yeast retrotransposon Ty3 employed for site-specific integration of heterologous genes with the advantages of stability and high copy number (Wang and Da Silva, 1996), or the filamentous fungus vector *Agrobacterium* able to transform the genera *Neurospora*, *Trichoderma*, *Aspergillus* and *Agaricus* (de Groot *et al.*, 1998).

3. Multicistronic expression vectors to allow one-step multigene metabolic engineering in mammalian cells (Fussenegger *et al.*, 1999).

4. Methods for stabilizing cloned genes, e.g. by integration into the chromosomes of host organisms.

5. Markers to search for and analyze metabolic pathways

The microbiological and analytical tools allow the evaluation of the effectiveness of the modified metabolic pathway:

1. Culture of the modified microorganism. In this respect, the ideal culture system enabling the application of mathematical and computational tools (see below) is continuous culture. However, it may happen that the stability of the genetically modified microorganism precludes the possibility of continuous maintenance of the culture in the long-term. In this case either fed-batch or batch systems have to be used.

2. Optimization of growth medium suitable for the operation of the desired metabolic pathway. For instance, an organism (*Serratia* spp.) modified with the bacterial hemoglobin gene (vgb) that is supposed to improve growth and to avoid by-product formation, has been reported to display various fermentation patterns according to the medium composition (Wei *et al.*, 1998).

3. Mass balance. This allows calculation of the yield of the desired product with respect to various substrates, and by comparison with maximal theoretical yields, evaluation of how far from the thermodynamic limit the metabolic pathway operates.

4. Isotopic labeling and analysis of blocked mutants as well as the determination of enzyme activities and metabolites, allow the determination of the effectiveness of operation of a given metabolic pathway contributing to the consumption of a certain substrate, or the formation of a required product.

5. The employment of non-invasive methods such as nuclear magnetic resonance and flow cytometry are preferred since they allow a direct evaluation of the performance of the microorganism under conditions similar to or identical with those in the industrial bioprocess.

With respect to the mathematical and computational tools the following considerations are important:

1. DNA data bases and software (Overbeek *et al.*, 2000; Covert *et al.*, 2001; see also below).

2. Metabolic pathways data bases, including kinetic and thermodynamic enzyme data. In this respect, several Internet sites are now available (Karp, 1998; Overbeek *et al.*, 2000; Covert *et al.*, 2001).

3. Tools designed for estimation of theoretical yields (e.g. from the metabolic pathway stoichiometry). This point is developed in Chapters 2 and 4.

4. Tools for the design of metabolic pathways. Several algorithms have been proposed for this purpose (Hatzimanikatis *et al.* 1996).

5. Quantitative tools for the simulation and prediction as well as analysis of the performance of the modified microorganism (e.g. Metabolic Control Analysis (MCA), Biochemical System Theory (BST) and Metabolic Flux Analysis (MFA)). These methods encompas a series of stoichiometric and linear

optimization procedures enabling the estimation of metabolic fluxes. Some of these tools are developed or shown in Chapters 2 and 4.

All these developments are integrated into the TransDisciplinary Approach (TDA) which is developed in the next section.

Engineering Cells for Specific Biotransformations

Possibilities for redirection of substrate fluxes either to microbial products or to biomass may be achieved either through modification of environmental parameters or by acting on the microorganism itself (Fig. 1.1) (Aon and Cortassa, 1997). Thus, we propose optimization of a specific biotransformation process by directed modification of the microorganism itself (rather than on that process) with the dual aim of achieving higher yields of products of economic interest and improved environmental quality.

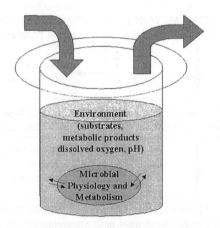

Figure 1.1. The microorganism as a target for bioengineering at metabolic, energetic, and physiological levels in chemostat cultures. The use of continuous cultures to study a microorganism at the steady state, provides a rigorous experimental approach for the quantitative evaluation of microbe's physiology and metabolism. Chemostat cultures allow the definition of the phase of behavior that suits the aim of the engineering, e.g. output fluxes of metabolic by-products of interest. Thus, it is a key tool of the TDA approach for the rational design of cells (see Chapters 4 and 5) whose flow diagram is shown in Fig. 1.2.

The originality of the present approach is that it integrates several disciplines into a coordinated scheme (i.e. microbial physiology and bioenergetics, thermodynamics and enzyme kinetics, biomathematics and biochemistry, genetics and molecular biology). Thus, it will be called a transdisciplinary approach (TDA). The TDA approach provides the basis for the rational design of microorganisms or cells in a way that has rarely been applied to its full extent. Progress in the area of a rational design of microorganisms has been hampered by the fact that few scientists can simultaneously master fermentation and recombinant DNA technologies along with mathematical modeling. In fact, in most cases researchers either apply sophisticated recombinant DNA technologies in a trial and error scheme or use mathematical techniques in isolation. The TDA approach can improve or optimize an existing process within an organism. The use of heterologous pathways for the production of new chemicals, or use various feedstock and cheap substrates or degrade xenobiotics, necessarily involves a previous modification of the organism using DNA recombinant techniques.

In general the TDA approach for MCE is iterative in nature, and may be outlined as follows (Fig. 1.2):

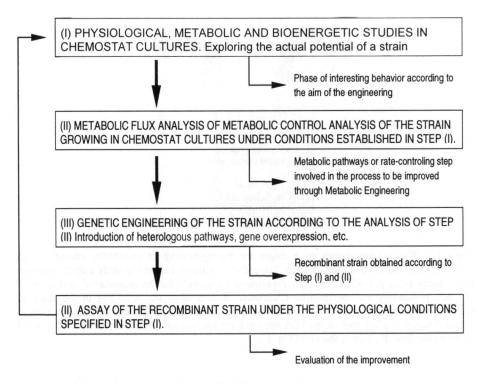

Figure 1.2. Engineering metabolic fluxes: The TDA approach.

(I) Physiological and bioenergetic studies are performed either in continuous, batch or fed-batch cultures according to the nature of the process.

The aim is to achieve a state known as "balanced growth" that allows the application of analytical tools such as MFA (see Chapter 2). Whenever a steady state is feasible, MCA can also be applied. Its expected outcome is determination of the most favorable behavior shown by the microorganism according to the aim of the engineering, e.g. ethanol production by *S. cerevisiae* at high growth rates in continuous cultures (see Chapter 4).

Although Step (I) is a clear one wherever we have a potential microorganism, several considerations must be taken into account. Often, continuous cultures cannot be run with cheap substrates (e.g. molasses, whey) especially in full-scale industrial processes. Under these conditions, batch or fed batch cultures must be used and mathematical modeling techniques such as MFA applied (see Step II).

The TDA approach can be initiated with a recombinant microorganism. Thereby it is possible to introduce heterologous metabolic pathways using DNA recombinant techniques to provide some microorganisms with novel activities, e.g. xylitol or arabinose degradation in *Zymomonas mobilis* (i.e. to introduce in *Z. mobilis* the ability to degrade xylose or arabinose). In this case, if chemostat cultures cannot be used (e.g. due to plasmid instability) then batch or fed-batch cultures may be employed to explore the recombinant's ability to perform the desired task. This organism may then be subjected to subsequent quantitative analysis.

(II) Metabolic studies facilitated by mathematical modeling. MCA and MFA of the strain performed under the conditions described in Step (I).

MFA may help determine the theoretical as well as the actual yields of the metabolite or macromolecule the production of which we seek to optimize. Moreover, the flux distribution and bioenergetic behavior may be investigated during the phase of interest (e.g. the growth rate at which the metabolite is maximally produced, excreted or accumulated intracellularly).

When applying the matrix form of MCA, the intracellular concentrations of the intermediates of the target pathway must be measured before further engineering. Enzyme kinetics must be investigated if information is not already available in the literature. The determination of the elasticity coefficients, their array in matrix form and the matrix inversion may be another aim of this step. Such inversion renders a matrix of control coefficients both for flux and metabolite concentrations. Thus, this step allows the identification of the rate-controlling steps of the flux or metabolite level in a metabolic pathway.

(III) Genetic engineering. Gene overexpression or up-modulation of the enzymes which control the flux. In the case of recombinant strains (e.g. those constructed by introduction of heterologous pathways) this step is still valid, since new rate-controlling steps of the specific biotransformation process may arise. The outcome of Step (III) is a modified microorganism optimized for a specific biotransformation process.

(IV) This step iterates Step (I): The engineered microorganism should be assayed under the physiological conditions defined in Step (I). The assay should allow evaluation of the improvement achieved in the biotransformation process with the use of the engineered microorganism.

We must stress that this rational approach is not followed in practice; instead we find examples with fragmentary applications of Steps (I)-(IV). As far as we are aware, the iteration between (IV)→(I) does not exist at present in the literature. What we find is a comparison between recombinant strains and their respective controls, i.e. the difference between them is used as a criterion of performance.

In Chapter 4 we describe in detail each of the steps as applied to examples with either prokaryotic or eukaryotic cells.

Metabolic Areas that Have Been Subjected to MCE

The main objective of this section is to update the work performed on MCE in the last years (Table 1.1), to highlight the main areas of metabolism that have been improved (Fig. 1.3), and to point out those areas to which less effort has been devoted.

MCE has dealt with manipulation of existing pathways or reactions aimed at producing a certain metabolite or macromolecule, or the introduction of new pathways or reactions into host cells. The main examples reported in Table 1.1 have been classified into five groups (Cameron and Tong, 1993; Stephanopoulos *et al.*, 1998; Lee and Papoutsakis, 1999): *(i)* enhanced production of metabolites and other biologicals already produced by the host organism; *(ii)* production of modified or new metabolites and other biologicals that are new to the host organism; *(iii)* broadening the substrate utilization range for cell growth and product formation; *(iv)* designing improved or new metabolic pathways for degradation of various chemicals especially xenobiotics; *(v)* modification of cell properties that facilitate bioprocessing (fermentation and/or product recovery).

The following pathways also shown in Fig. 1.3 have been subjected to intense MCE: Pyruvate pathway for production of organic acids or flavor compounds in bacteria such as acetoin and diacetyl (Platteeuw *et al.*, 1995; Lopez de Felipe *et al.*, 1998; see also Chapter 4); sugar transport (hexose transporters: see Özcan and Johnston, 1999, for a review; phosphotransferase system (PTS):

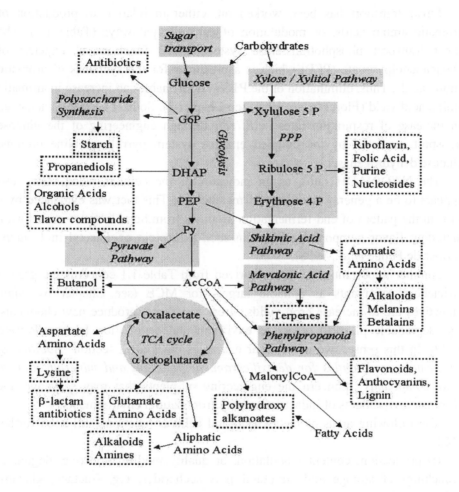

Figure 1.3. Primary and secondary metabolic routes and their connections to central catabolic pathways through common metabolites. Schematic representation of some pieces of information detailed in Table 1.1. On grey boxes are represented multiple step pathways while end products are enclosed in dashed-rectangles. Several of the pathways shown have been subjected to intensive Metabolic Engineering. The pathways depicted include those described for plants, bacteria and fungi. The pathways for plants were taken from Morgan *et al.* (1999).

Liao *et al.*, 1996; Gosset *et al.*, 1995), arabinose and xylose assimilating pathways (Zhang *et al.*, 1995; Deanda *et al.*, 1996; see Chapter 4); pathways for production of propanediol (Cameron *et al.*, 1998), butanol (Papoutsakis and Bennett, 1999), lysine (Vallino and Stephanopoulos, 1993; Eggeling and Sahm, 1999), aromatic aminoacids (Gosset *et al.*, 1995; Liao *et al.*, 1996; Flores *et al.*, 1996), β-lactam antibiotics (Khetan and Hu, 1999).

Sugar transport has been worked on, either in relation to production of aromatic amino acids, or modulation of catabolic pathways (Table 1.1). The sugar transport phosphotransferase system (PTS) functions at expense of phosphoenolpyruvate (PEP) which is a precursor for the synthesis of aromatic amino acids. Thus, elimination of the PTS system enables an increase in aromatic amino acid yield (Flores *et al.*, 1996, see Chapter 3). Sode *et al.* (1995) achieved an increase of respiration rates in *E. coli* through engineering of the glucose transport mediated by the coupled enzyme system pyrroloquinoline quinone glucose dehydrogenase.

The NAD/NADH ratio, as an indicator of the cellular redox state, also appears to be a general modulator of metabolism. This fact was evidenced by a shift in the pattern of end-fermentation products from homolactic to mixed acid, including flavor compound production such as acetoin or diacetyl in *L. lactis* (Lopez de Felipe *et al.*, 1998).

Another important observation derived from Table 1.1 concerns the growth efficiencies of plants as suitable targets for MCE (see Chapter 6). Main objectives are to increase crop yields, stress resistance, produce new chemicals, e.g. degradable plastics, or act as environmental depollution factors (Poirier, 1999). In this sense, two of the four examples covered in section "*Addition of new catabolic activities for detoxification, degradation and mineralization of toxic compounds*", concern the engineering of plants that may scavenge and eliminate by-products of industrial activity from natural environments.

The following areas of metabolism and cellular functions still need further MCE:

(i) production, content modulation, or quality modification (e.g. degree of branching) of storage and structural polysaccharides, e.g. fructans, glucans, xylans, pectins.

(ii) cellular transport functions (e.g. of protons, sodium, calcium).

(iii) cellular processes (e.g. cell division cycle, catabolite repression, cell differentiation).

Table 1.1 Examples of metabolic engineering developments in main areas of biotechnological interest

Improved production of chemicals and proteins produced by the host organism

Chemical	Host organism	Notes (pathway engineered or modification method applied)	Reference
Xylanase	*Streptomyces lividans*	The signal peptide of xylanase A (XlnA) was replaced with signal peptides of mannanase A abd cellulase A increasing the expression level 1.5–2.5 fold, according to the length of the signal peptide sequence	Pagé *et al.*, 1996
Aromatic compounds	*Escherichia coli* non PTS glucose transporting mutants	*E. coli* with an inactivated PTS system with Gal permease, glucokinase, to internalize and phosphorylate glucose, plus the tklA coding for transketolase allowed higher yields of aromatic compounds	Flores *et al.*, 1996
Riboflavin, folic acid and purine nucleosides	*Bacillus subtilis*	A diagnosis of the controling steps driving carbon to the synthesis of purine related compounds revealed that the central amphibolic pathways supplying carbon intermediates and reduced cofactors, are not limiting	Sauer *et al.*, 1997, 1998
Organic acids	*Saccharomyces cerevisiae*	Overexpression of malate dehydrogenase result in accumulation of malic, citric and fumaric acids revealing pyruvate carboxylase as a limiting factor for malic acid production	Pines *et al.*, 1997
Diacetyl and/or acetoin	*Lactococcus lactis*	Expression of *Streptococcus* mutans NADH oxidase that decreased the NADH / NAD ratio resulted in a shift from homolactic to mixed acid fermentation with the activation of the acetolactate synthase and acetoin or diacetyl production.	Lopez de Felipe *et al.*, 1998
Cellulose accumulation and decreased lignin synthesis	Aspen (*Populus tremuloides*)	The lignin biosynthetic pathway was down-regulated by antisense mRNA expression of 4-coumarate coenzyme A ligase. Trees exhibited 45% decrease in lignin and a compensatory 15% increase in cellulose	Hu *et al.*, 1999
Lactic acid	*Escherichia coli*	Overexpression of homologous and heterologous phosphofructokinase and pyruvate kinase resulted in altered fermentation patterns: increase in lactic acid with decreased ethanol production	Emmerling *et al.*, 1999

Table 1.1 (*Continued*)

Cephalosporin	*Streptomyces clavuligerus*	Kinetic analysis and modeling indicated that the rate controlling step was the lysine-epsilon-aminotransferase that when overexpressed exhibited higher yield of the β lactam antibiotic	Khetan *et al.*, 1996
Carbon dioxide	*Saccharomyces cerevisiae*	Overexpression of the enzymes catalyzing the lower glycolytic pathway led to increased rates of CO_2 production but not to increased ethanol production only during glucose pulses.	Smits *et al.*, 2000
Pyruvate	*Torulopsis glabrata*	Metabolic flux analysis was used to point the most appropriate culture conditions regarding glucose, dissolved oxygen and vitamin concentrations to optimize the production of pyruvate under fed-batch cultivation	Hua *et al.*, 2001

Production of chemicals and proteins new to the host organism

Chemical	Host organism	Notes (pathway engineered or modification method applied)	Reference
Carotenoids	*Candida utilis*	Yeast cells transformed with the ctr operon from *Erwinia* coding for enzymes of carotenoid synthesis, redirect farnesyl pyrophosphate, an intermediate in the synthesis of ergosterol toward lycopene and β-carotene	Shimada *et al.*, 1998
Lycopene	*Escherichia coli*	Expression of the Pps (phosphoenolpyruvate synthase) and deletion of the PYK (pyruvate kinase) activities result in higher yields of lycopene production. This indicates that the isoprenoid synthesis in *E. coli* is limited by the availability of glyceraldehyde3P	Farmer and Liao (2001)
Pregnenolone and progesterone	Yeast	Introduction of adrenodoxin (ADX), and adrenodoxin reductase (ADR) P450scc, β OH steroid dehydrogenase-isomerase; Delta7 sterol reductase and disruption of delta22 sterol desaturase from the steroidogenic pathway from ergosterol	Duport *et al.*, 1998
Low molecular weight fructan	Sugar beet	Transformation with a sucrose fructosyl transferase from *Helianthus tuberosus*	Sevenier *et al.*, 1998

Table 1.1 (*Continued*)

Globotriose and UDP galactose	Recombinant *E. coli* and *Corynebacterium ammoniagenes*	*E. coli* cells overexpressing UDP-Gal biosynthetic genes with *C. ammoniagenes* able to produce globotriose from orotic acid and galactose	Koizumi *et al.*, 1998
Gamma linolenic acid	Tobacco plants	Transformation of tobacco plants with a cyanobacterial delta 6 desaturase that allow to change the composition of fatty acids introducing polyunsaturated ones.	Reddy and Thomas, 1996
Medium chain length poly-hydroxyalkanoates	*Arabidopsis thaliana*	β-oxidation of plant fatty acids generate various R-3-hydroxyacyl-CoA that serve as precursors of polyhydroxyalkanoates in plants transformed with polyhydroxyalkanoate synthase (PhaC1) synthase from *Pseudomonas aeruginosa* and the gene products were directed to peroxisomes and glyoxysomes.	Mittendorf *et al.*, 1998
L-alanin	*Lactococcus lactis*	*Bacillus sphaericus* alanine dehydrogenase was introduced into a *L. lactis* strain deficient in lactate dehydrogenase shifting the carbon fate from lactate toward alanine.	Hols *et al.*, 1999
1,2 Propanediol	*Escherichia coli*	NADH-linked glycerol dehydrogenase expressed together with methylglyoxal synthase improved up to 1.2 g/l.	Altaras and Cameron, 1999
Isoprenoid	*Escherichia coli*	Isopentenyl diphosphate isomerase gene from yeast overexpressed in an *E. coli* strain that expressed *Erwinia* carotenoid biosynthetic genes resulted in increased accumulation of beta carotene.	Kajiwara *et al.*, 1997

Extension of substrate range for growth and product formation

Substrate	Host organism	Notes (pathway engineered or modification method applied)	Reference
Pentoses such as xylose and arabinose	*Tetragenococcus halophila*	Mannose PTS (PEP:mannose phosphotransferase, PFK and glucokinase triple mutants are able to ferment pentoses in the presence of hexoses).	Abe and Higuchi, 1998
Xylose	*Saccharomyces cerevisiae*	Xylulokinase gene was introduced in an recombinant yeast strain that expresses the xylose reductase and xylitol dehydrogenase. The resulting strain is able to perform ethanolic fermentation from xylose as sole carbon source at high aeration levels	Toivari *et al.*, 2001

Table 1.1 (*Continued*)

| Starch | *Saccharomyces cerevisiae* | Yeast cells expressing an active glucoamylase from *Rhizopus oryzae* were able to use starch as C-source. The foreign protein was targeted to the cell wall through fusion with a yeast α-agglutinin. | Murai *et al.*, 1997 |
| Arabinose fermentation | *Zymomonas mobilis* | Arabinose isomerase, ribulokinase, ribulose 5 phosphate epimerase, transaldolase and transketolase from *E. coli* were introduced into *Z. mobilis* under the control of a constitutive promoter | Deanda *et al.*, 1996 |

Addition of new catabolic activities for detoxification, degradation and mineralization of toxic compounds

Chemical	Organism	Notes (source and type of bioremediation genes)	Reference
Mercury (II)	Yellow poplar engineered with the mercury reductase (MerA) gene	Converts the highly toxic Hg(II) to Hg(0) by transformation with a bacterial MerA gene.	Rugh *et al.*, 1998
Organopollutant degradation (toluene and trichloroethylene) and heavy metals arsenic, chromium, lead, cesium plutonium and uranium.	*Deinococcus radiodurans*	TOD genes to degrade organopollutants (toluene deoxygenase, a flavoprotein, a ferredoxin, and a terminal oxygenase genes) expressed in radioactive environments.	Lange *et al.*, 1998
Mercury (II)	*Deinococcus radiodurans*	The mercury resistance operon mer from *E.coli* were introduced in a highly radiation resistant bacterium with the aim of remediating radioactive waste contaminated with heavy metals. The mercury resistance levels correlated with the gene dose of the integrated operon	Brim *et al.*, 2000
Explosives	Tobacco plants	Transgenic plants expressing pentaerythritol tetranitrate reductase from *Enterobacter cloacae* are able to degrade glycerol trinitrate and potentially trinitrotoluene, pollutants commonly present in military sites.	French *et al.*, 1999b
Biphenyls	*Pseudomonads*	The biphenyl degradative pathway was introduced into *Pseudomonads* living in the rhizosphere with the potential use to bioremediate polluted soil.	Brazil *et al.*, 1995

Table 1.1 (*Continued*)

Modification of cell properties

Property	Organism	Notes (strategy and rationale of modification)	Reference
Alteration of source–sink relations and carbon partitioning provoking direct effect on growth of plants	Tobacco plants	Plants tranfected with the yeast invertase gene under the control of an ethanol inducible promoter avoid the deleterious effects on growth of a constitutive expression.	Caddick *et al.*, 1998
Decrease of O_2 photosynthesis inhibition	Rice plants	Maize phosphoenolpyruvate carboxylase introduced into C3 plant enhances photosynthesis by acquisition of part of the metabolic machinery to concentrate CO2 characteristic of C4 plants (maize).	Ku *et al.*, 1999
Herbicide resistance (glyphosate)	Tobacco cells	Introduction of the EPSPS from petunia into the chloroplast genome avoids escape and dissemination of the foreign genes because of their absence into pollen cells	Daniell *et al.*, 1998
Harvest index	*Nicotiana tabacum*	Phytochrome A (heterologous (oat) apoproteins) gene was introduced into tobacco under the control of 35S CaMV promoter.	Robson *et al.*, 1996
Low temperature resistance of higher plants	Tobacco plants	Delta 9 desaturase from cyanobacteria introduced in tobacco exhibit reduced levels of saturated fatty acids in membrane lipids and increased chilling resistance.	Ishizaki-Nishizawa *et al.*, 1996
Altered fermentation pattern with production of 2,3-butanediol and acetoin	*Serratia marcescens*	*Serratia marcescens* transformed with the bacterial (*Vitreoscilla*) hemoglobin gene (vgb) where growth is not necessarily improved, but fermentation pattern altered according to medium composition.	Wei *et al.*, 1998
Enhanced growth and altered metabolite profiles	*Nicotiana tabacum*	Transgenic tobacco plants expressing Vitreoscilla hemoglobin gene exhibit better yield and faster growth and altered alkaloids contents (nicotine and anabasine).	Holmberg *et al.*, 1997
Enhanced potato tuber growth with altered sugar content	Potato plants	Yeast invertase expression either in the apoplast (extracellular space) or cytoplasm, allow increased or decreased, tuber size accompanied with lower or higher, tuber numbers per plant, respectively.	Sonnewald *et al.*, 1997

Table 1.1 (*Continued*)

Altered fermentation pattern: shift from a homolactic to a mixed acid fermentation by perturbing the redox status of the cell.	*Lactococcus lactis*	Transformation with the *Streptococcus* NADH oxidase (nox2) gene resulting in a mixed acid fermentation according to the expression level of NADH oxidase and to the associated redox status of the cell.	Lopez de Felipe *et al.*, 1998
Hybridoma cells growth in glutamine-free media	Hybridoma cells	Glutamine synthetase from Chinese hamster was introduced into a hybridoma cell line achieving parental levels of antibody production in a glutamine-free medium	Bell *et al.*, 1995
Decreased yield and growth rate	*Escherichia coli*	Decreasing the number of lipoyl domains per lipoate acetyltransferase in pyruvate dehydrogenase resulted in adverse effects on growth and biomass yield	Dave *et al.*, 1995
Expression of *E. coli* glycine betaine synthetic pathway and yeast trehalose synthetic genes	Plants	Transgenic plants transformed with genes from choline-to-glycine betaine pathway enhances stress tolerance to cold and salt, whereas those carrying yeast TPS1 (UDP dependent trehalose synthesis pathway) exhibit higher draught tolerance but with negative side effects.	Strom, 1998
Enhanced recombinant protein production	*Escherichia coli*	Acetolactate synthase from *Bacillus* was introduced in *E. coli* to drive the excess pyruvate from the glycolytic flux away from acetate to acetolactate which was then converted into acetoin, a less toxic metabolite from the point of view of heterologous protein production.	Aristidou *et al.*, 1995
Suppressed acid formation	*Bacillus subtilis*	Mixed substrate consumption, glucose and citrate, the latter exerting likely regulatory roles on glycolytic enzymes, PFK and PK.	Goel *et al.*, 1995
Anaerobic growth and improved ethanol yield	*Pichia stipitis*	The *S. cerevisiae* URA1 gene encodes a dihydroorotate dehydrogenase that uses fumarate as an alternative electron acceptor enabling anaerobic growth and fermentation in *P. stipitis* when transformed with this gene.	Shi and Jeffries, 1998
Changes in sugar utilization pattern	*Escherichia coli*	Overexpression of pyrroloquinoline quinone glucose dehydrogenase resulted in increased sugar-dependent respiration according to the quality of the carbon source (either PTS or non PTS sugar)	Sode *et al.*, 1995

From DNA Sequence to Biological Function

In the last half century we have witnessed outstanding scientific discoveries and technological achievements: *(i)* the elucidation of the mechanism whereby genetic information is encoded, processed and expressed, *(ii)* understanding of the physico-chemical basis of self-organization in artificial and biological systems, *(iii)* the detailed development of cellular biology and its dynamic quantification by use of fluorescence microscopy, image digitization and processing, *(iv)* the accumulation and systematization of genetic and biochemical information, and its global availability through the Internet.

In the meantime it has become increasingly clear that the availability of integrated databases, even with the growing computer power for their fast analyzes, will not in itself be sufficient to take us from DNA sequence to biological function. Basically, this is because expression of the phenotype depends on the spatio-temporal display of genetic information through nonlinear mass-energy-information-carrying networks of reactions. Two main characteristics complicate the generation of biological organization (Lloyd, 1992, 1998; Aon and Cortassa, 1997): *(i)* simultaneous expression on multiple levels of organization spanning a broad range of spatio-temporal scales; *(ii)* the ability to generate emergent phenomena, visualized as spatio-temporal structures, of transient or enduring nature, that determine the system's structure and dynamic behavior.

Table 1.2 (Miklos and Rubin, 1996) reminds us that a unicellular protozoan, a nematode worm, and a fly, develop and function with 12,000–14,000 genes. It is thought that the four to six times as many genes that the humans have over those of *Caenorhabditis* and *Drosophila*, may have occurred by polyploidization, a common evolutionary feature in most unicellular and metazoan lineages (Miklos and Rubin, 1996). If the genome projects verify the underlying octoploid nature of the human and mouse genomes, then the basic vertebrate gene number may be similar to that of the fly and worm, about 12,000 to 14,000 genes. These examples illustrate that there can be large differences in morphological complexity among different organisms that have similar numbers of genes (Miklos and Rubin, 1996). Thus, gene number *per se* is not likely to provide a useful measure of biological complexity.

As a result of a multiplicity of regulatory interactions at all levels in cells, a change at one level in the complex network does not necessarily lead to a particular change in function or phenotype (Harold, 1990; Aon and Cortassa, 1997; Fiehn *et al.*, 2000). Morphology and organization, like other complex phenotypes, are the outcome of interactions among the expression of multiple genes and the environment. The relationship of genes to cell form and function is not like that of genes to proteins. Due to phenotypic complexity, enzyme activity and fitness are not unambiguously related (Dykhuizen *et al.*, 1987; Harold, 1990; Aon and Cortassa, 1997). The so called "functional genomics" seeks to unravel the role of unknown genes as well as, in general, to move from DNA sequence to biological function (Oliver, 1996; Fiehn *et al.*, 2000; Raamsdonk *et al.*, 2001). For functional analysis different methods are employed at the level of the genome, transcriptome, proteome, or metabolome (Oliver, 2000); the main difference between them being that the last three levels are context-dependent, i.e. the entire complement of messenger RNA (mRNA) molecules, proteins or metabolites in a tissue, organ or organism varies with physiological, pathological or developmental conditions (Oliver, 2000). Unlike mRNA molecules, proteins and metabolites are functional entities within the cell. For *S. cerevisiae* there are fewer than 600 low-molecular-weight intermediates, whereas there are *ca.* 6,000 protein encoding genes (Raamsdonk *et al.*, 2001). This means that there is no direct relationship between metabolite and gene as there is for mRNAs and proteins. However, mutations producing no overt phenotype can still produce changes in the concentration of intracellular metabolites. Thus, mutants that are silent when scored on the basis of metabolic fluxes, may still produce effects on metabolite concentrations. This idea of using analysis of metabolites (also called "metabolite profiling") is being exploited to reveal the phenotype of silent mutations in yeast (Raamsdonk *et al.*, 2001) or plant functional genomics (Fiehn *et al.*, 2000).

The phenotypic consequences of gene inactivation depend on genetic background and pleiotropic effects (see Chapter 7). The latter implies on the one hand, that a gene knockout can result in different phenotypes when it performs in different genetic backgrounds (Miklos and Rubin, 1996), and on the other hand that specific environmental conditions may decide when the knockout of a certain gene may be lethal, or the effect of a mutation manifested. Most of the yeast genome encodes proteins that are synthesized during vegetative growth. Single gene disruption in the case of many yeast genes does not result in an obvious growth defect (Burns *et al.*, 1994). Therefore, a large fraction of the genes that determine a growth defect can only be uncovered by testing a variety of different growth conditions. In yeast, for example, the total deletion of a membrane protein coding for a putative acetic acid exit pump usually has little phenotypic effect. However, the cells die when grown with glucose at low pH or when perturbed with acetic acid (Oliver, 1996). Another example in yeast is that the temperature-sensitive cell division cycle gene *CDC28* does not express its temperature sensitivity when the mutant yeast cells are grown at the restrictive temperature in the presence of gluconeogenic substrates (Mónaco *et al.*, 1995; Aon and Cortassa, 1995). This effect was completely reversible on glucose addition (Aon and Cortassa, 1995, 1997).

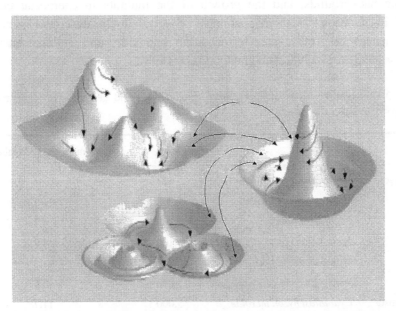

Figure 1.4. A geometric interpretation of *homeodynamics*. Several types of attractors with their corresponding basins of attraction are represented. The putative trajectories followed by system's dynamics are emphasized by arrows as well as the separatrices between basins. The *homeodynamic condition* implies that the system's dynamics visualized as a fluid flowing around itself, may shift between attractors at bifurcation points where stability is lost. Thereby, the system's dynamics following a perturbation, flies away toward another attractor exhibiting either qualitative or quantitative changes in its behavior. The upper left 3-dimensional (3D) plot shows saddle and fixed points; the latter with different values, each one representing a different branch of steady states. Alternative occupancies of these states, following the change of a bifurcation parameter, gives a bistable switch with memory-like features. Also stable and unstable foci are depicted in the upper left 3D plot. The lower right 3D plot, shows a limit cycle with its basin of attraction that may be attained through an unstable focus, characteristic of oscillatory behavior (self-sustained or damped oscillations, respectively). The middle 3D plot depicts an attractor with three orbits embedded in it, with the potential for chaotic behavior (Reproduced from Lloyd, Aon and Cortassa, 2001. *TheScientificWorld* 1, 133-145).

Pleiotropy can arise if a protein is functionally required in different places, or at different times, or both (Miklos and Rubin, 1996). Pleiotropic effects may also arise through the strategic function of a gene product deeply nested in metabolic or regulatory networks which might affect several processes either simultaneously or in sequence. For example, in yeast, the products (proteins) of

*SNF*1 or *SNF*4 genes have been postulated to be involved in a regulatory network which triggers the derepression of several gluconeogenic enzymes when yeast grows in the presence of non-fermentable carbon sources e.g. ethanol, acetate or glycerol (Schuller and Entian, 1987). The deletion of *SNF*1 or *SNF*4 genes within isogenic backgrounds, and the growth of the mutants in chemostat cultures, reveal in addition to their postulated effects, newly described pleiotropic consequences on cell cycle, fermentative behavior, and cellular energetics (Cortassa and Aon, 1998, Aon and Cortassa, 1998).

Table 1.2 Current predictions of approximate gene number and genome size in organisms in different evolutionary lineages

		Genes	Genome Size in Megabases
Prokaryota	*Mycoplasma genitalium*	473	0.58
	Haemophilus influenzae	1,760	1.83
	Bacillus subtilis	3,700	4.2
	Escherichia coli	4,100	4.7
	Myxococcus xanthus	8,000	9.45
Fungi	*Saccharomyces cerevisiae*	5,800	13.5
Protoctista	*Cyanidioschyzon merolae*	5,000	11.7
	Oxytricha similis	12,000	600
Arthropoda	*Drosophila melanogaster*	12,000	165
Nematoda	*Caenorhabditis elegans*	14,000	100
Mollusca	*Loligo pealii*	>35,000	2,700
Chordata	*Ciona intestinalis*	N	165
	Fugu rubripes	70,000	400
	Danio rerio	N	1,900
	Mus musculus	70,000	3,300
	Homo sapiens	70,000	3,300
Plantae.	*Nicotiana tabacum*	43,000	4,500
	Arabidopsis thaliana	16,000–33,000	70–145

N, not available.
Reprinted from *Cell*, 86, Miklos and Rubin, The role of the genome project in determining gene function: Insights from model organisms, 521-529. ©copyright 1996, with permission from Elsevier Science.

Temporal and Spatial Scaling in Cellular Processes

Biological processes at sub-cellular, cellular and supra-cellular levels scale in space and time (Lloyd, 1992; Aon and Cortassa, 1993, 1997). These phenomena seem to arise from the complexity of living systems with respect to their simultaneous structural and functional organization at many levels of organization. It has been suggested that the emergent property of cell function arises from transitions between levels of organization at bifurcation points in the dynamics of biological processes (Aon and Cortassa, 1997). At bifurcation points, a dynamic system loses stability and behavioral changes occur. These may be quantitative, qualitative, or both (Fig. 1.4). Quantitatively, it may happen that the system dynamics moves at a limit point to a different branch of steady state behavior, (lower or higher), e.g. as in bistability. Under these conditions, the system does not change its qualitative behavior, i.e. it continues to be at a point attractor, either a stable node or a focus. However, at some bifurcation points, drastic qualitative changes occur; the system evolves from a monotonic operation mode toward periodic, (Hopf bifurcation), or chaotic, motions (Nicolis and Prigogine, 1977, 1989; Abraham, 1987; Aon *et al.*, 1991). Thus, dynamically-organized phenomena are *homeodynamic* (Fig. 1.4), and are visualized as demonstrating spatio-temporal coherence. Under *homeodynamic* conditions a system, (e.g. network of reactions or cells), may exhibit emergent spatio-temporal coherence, i.e. dynamic organization (Lloyd *et al.*, 2001). A graphic analogy of the concept of *dynamic organization* under *homeodynamic* conditions is shown in Fig. 1.5.

Dynamic organization in cells or tissues, is thus an emergent property arising from transitions between levels of organization at bifurcation points in the dynamics of biological processes. In Fig. 1.5 the different landscapes represent the dynamic trajectories of sub-cellular processes (e.g. enzyme activity, synthesis of macromolecules, cell division; all indicated as spheres in the plot), resulting from the functioning of those processes on different spatio-temporal scales, (levels of organization). The dotted lines that link the spheres, (different sub-cellular processes), indicate the coupling between them. The coupling between processes that function simultaneously on different spatio-temporal scales, *homeodynamically* modifies the system trajectories (the landscapes' shape), as represented by the movement of the spheres through peaks, slopes and valleys. The sphere on the landscape on top symbolizes a process occurring at a higher level of organization, (higher spatial dimensions and lower relaxation times), i.e. a macroscopic spatial structure, (waves, macromolecular networks, subcellular organelles, etc.). Indeed, the functioning of the system is coordinated and coupling occurs top-bottom as well as bottom-up. The interdependent and coupled cross-talk between the two flows of information crosses levels of organization through and beyond each level (Lloyd *et al.*, 2001).

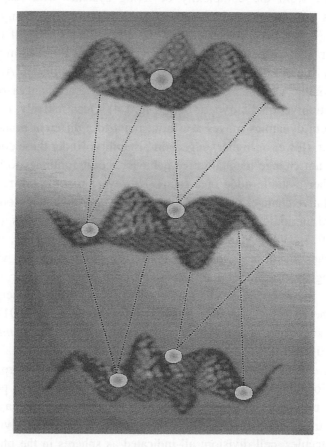

Figure 1.5. A graphic analogy of the concept of *dynamic organization* under *homeodynamic* conditions. *Dynamic organization* in, e.g. cells or tissues, is an emergent property arising from transitions between levels of organization at bifurcation points in the dynamics of biological processes (Reproduced from Lloyd, Aon and Cortassa, 2001. *TheScientificWorld* 1, 133–145).

Scaling in Microbial and Biochemical Systems

The spatio-temporal scaling shown by sub- and supra-cellular processes allows interpretation of the balanced growth exhibited by microorganisms. Any two state variables of an exponentially growing microbial system are related by an allometric law (Rosen, 1967, 1970; Aon and Cortassa, 1997). In the case of balanced growth these two state variables may represent N, a population of microbes (biomass), and M, a population of macromolecules (protein, carbohydrate) synthesized in a constant proportion with the microbial biomass. Growth is balanced when the specific rate of change of all metabolic variables (concentration or total mass) is constant (see Eq. 3.1) (Barford *et al.*, 1982; Roels, 1983; Cooper, 1991; Cortassa *et al.*, 1995):

Hence the fact that several sub- and supra-cellular systems scale their functioning in space and time exponentially, suggests that: *(i)* a defined relationship exists between the whole (cell) and each of its constituents (macromolecules); *(ii)* processes happen over broad time scales, given by their relaxation times following perturbation (see below).

Temporal scaling in microbial and biochemical systems implies that these react with different relaxation times toward a perturbation. We prefer the term scaling to hierarchy because in biological systems in general, and cellular ones in particular, there are vertical flows of information which coexist with horizontal ones. These mutual and reciprocal interactions are effected through coupling of sequential and parallel processes, respectively. This is clearly the case for a spatially highly interconnected and dynamically coupled system with massive occurrence of sequential and parallel processing such as a cell.

The consequences of the temporal scaling for cell function are diverse and important:

(i) In coupled processes, a variable in one dynamic subsystem because of its fast relaxation toward perturbations may act as a parameter of another dynamic subsystem that relaxes slowly (Bertalanffy, 1950; Aon and Cortassa, 1997).

(ii) Significant simplifications through reduction in the number of variables may be achieved by analysis of relaxation times. Essentially, the system's description is reduced to the slow variables (i.e. long relaxation times) (Heinrich *et al.*, 1977; Reich and Sel'kov, 1981; Roels, 1983). Otherwise stated, the overall dynamics of a system is governed by the relaxation times of the slow processes even though the system contains rapid motions.

(iii) By comparing the relaxation times of the intracellular processes with respect to those characterizing relevant changes in environmental conditions, it may help to decide whether the changes in the environment occur much faster than the mechanism by which the organism is able to adjust its activities or vice versa (Esener *et al.*, 1983; Roels, 1983; Vaseghi *et al.*, 1999). As an example, the concentrations of ATP and NADH exhibit rapid relaxation times because of their high turnover rates. In other words, the rate of adaptation of the mechanisms involved in keeping the intracellular concentrations of ATP and NADH is rapid

as compared with the characteristic times of the changes in substrate concentration in the environment. Hence, the system's behavior can be directly expressed in terms of the substrate concentration in the environment; this is a consequence of a pseudo-steady state assumption (Roels, 1983).

(iv) Temporal scaling allows us to understand that some cellular process (e.g. a metabolic pathway) may achieve a balanced growth condition before others. Thus, it has been suggested that a microorganism may apportion its total energy-producing capacity more rapidly between fermentation and respiration long before adjustment of the specific rates of oxygen uptake or carbon dioxide production (Barford *et al.*, 1982). Thus, the respiratory quotient attains a balanced condition before either of its components.

Views of the Cell

Black and Grey Boxes: Levels of Description of Metabolic Behavior in Microorganisms

The chemical reaction equation for the synthesis of a microorganism is a complex one. The following chemical equation represents the amount of carbon, NADPH, NADH, NH_4^+ and CO_2 required or produced during the synthesis of one gram of yeast cell biomass from glucose as carbon source :

$$7.4\ C_6H_{12}O_6 + 7.2\ NH_3 + 7.9\ NADPH + 14.5\ NAD^+ \longrightarrow$$
$$9.9\ C_4H_{7.5}O_{1.7}N_{0.73} + 4.7\ CO_2 + 7.9\ NADP + 14.5\ NADH + 18\ H_2O$$

The arrow which indicates the direction of the reaction "hides" a complex network of around one thousand chemical reactions occuring inside an organism or cell, e.g. in bacteria (Bailey and Ollis, 1977; Stouthamer and van Verseveld, 1987; Stephanopoulos and Vallino, 1991; Varma and Palsson, 1994), yeast (Cortassa *et al.*, 1995; Aon and Cortassa, 1997; Vanrolleghem *et al.*, 1996), or mammalian cells (Zupke and Stephanopoulos, 1995; Vriezen and van Dijken, 1998).

The description of a complex network of chemical reactions may be achieved at different levels of explanation or detail. Schematically, there are two levels of description: black and grey boxes. In the black box approach only the input(s) and output(s) of the system (e.g. microorganisms) are specified (see Chapters 2 and 3). If we progressively improve our descriptive ability by increasing the knowledge about what is occurring inside the box (e.g. mechanisms, reactions), it then becomes a grey box (see Chapter 4). The grey level of the box will be darker or lighter, depending on how deep is our knowledge of the physiological and dynamic conditions of the system under study (see Chapter 5).

The complex multilayered cellular circuitry shown in Fig. 1.6 intends to depict the complexity that we face in MCE. On the one hand, each line connecting two nodes (metabolites) of the metabolic network (bottom layer) is catalyzed by an enzyme, the amount of which is defined by gene expression (top layer) (transcription, translation, post-translational modifications). The enzyme activity (rate at which the conversion of one intermediate into another proceeds in the bottom layer) will be determined either by the intrinsic reaction dynamics, e.g. substrate inhibition, product activation (bottom layer) or intracellular signalling pathways, e.g. allosteric regulation, or covalent modification (top layer).

A typical mammalian cell synthesizes more than 10,000 different proteins, a major proportion of which are enzymes that carry out the mass-energy transformations of the network shown in the bottom layer. The information-processing function of cells is overseen by other networks (top layer), i.e. gene expression and intracellular signalling pathways, the latter intertwined with the bottom layer (Fig. 1.6). Many proteins in living cells function primarily as transfer and information-processing units (Bray, 1995). At least one third of cellular proteins take part in cellular macromolecular networks (e.g. cytoskeleton) and laminae of the cytoplasm and nucleus (Penman *et al.*, 1981). Compelling experimental evidence suggests that metabolism is strongly associated with cellular scaffolds, and that these insoluble matrices and their dynamics in turn deeply influence the dynamics of chemical reactions, i.e. they also belong to the top layer (Clegg, 1984, 1991; Cortassa and Aon, 1996; Aon and Cortassa, 1997, Ovadi and Srere, 2000, and Saks *et al.*, 2001; Aon *et al.*, 2000, 2001, for reviews). The latter emphasizes the multidimensional character of physiological responses in cells. The discovery of signaling molecules that interact with microtubules as well as the multiple effects on signaling pathways of drugs that destabilize or hyperstabilize microtubules, indicate that cytoskeleton polymers are likely to be critical to the spatial organization of signal transduction (Gundersen and Cook, 1999).

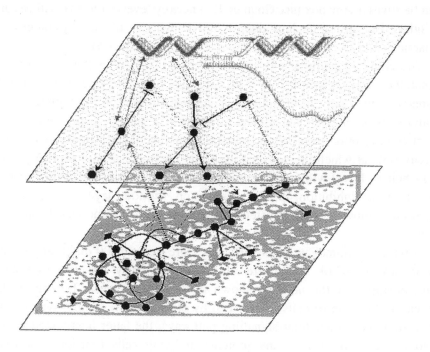

Figure 1.6. A view of cells as multilayered mass-energy-information networks of reactions. Metabolic reactions embedded in the cytoplasmic scaffolds are shown diagrammatically with each chemical species represented by a filled circle (bottom layer). Central catabolic pathways (glycolysis and the TCA cycle) are sketched. The chemical reaction network for the synthesis of a microorganism, (e.g. bacteria, unicellular fungi), is a complex one comprising around 1,000 chemical reactions (Alberts *et al.*, 1989) (see text for further explanation). A typical mammalian cell synthesizes more than 10,000 different proteins, a major proportion of which are enzymes that carry out the mass-energy transformations of the bottom-layer network. The information-processing function of cells is carried out by another networks, i.e. intracellular signalling pathways and the regulatory circuitry related with gene expression (top layer), intertwined with the network depicted in the bottom layer. The information-carrying networks are shown in a different layer just for the sake of clarity and its presence on top does not imply hierarchy. On the contrary, cross-talk connections existing between the two layers are emphasized by arrows (activatory, with arrowheads, or inhibitory, with a crossed line). Nevertheless, each network has its own type and mechanisms of interactions, e.g. the nodes on the top layer are proteins or second messengers. The proteins are continuously synthesized through transcription, translation mechanisms of gene expression, and exert feed-back regulation (e.g. DNA-binding proteins) on its own or the expression of other proteins, e.g. enzymes (depicted as double arrows on top layer). Moreover, the proteins taking part of cytoskeleton either exert feed-back regulation on its own expression (e.g. modulation of intracellular levels of tubulin through the stability of mRNAs) (Cleveland, 1988), or influence metabolic fluxes through epigenetic mechanisms (e.g. (de)polymerization, dynamic instability) (see Aon and Cortassa, 1997; Aon *et al.*, 2000a,b, for reviews; Lloyd *et al.*, 2001) (see text for further explanation).

Together with the metabolic and signaling networks shown in Fig. 1.6, we must consider the regulatory circuitry required for gene expression, involving transcriptional activators, suppresors, *cis* or *trans*-acting factors, most of them consisting of proteins, or DNA-binding proteins (Fig. 1.6, top layer). This regulatory net mainly determines the transcriptional level of gene expression.Often these transcriptional regulatory schemes are deduced from qualitative studies of the molecular biology of recombinant microorganisms (e.g. bacteria, yeast) grown on agar plates. These recombinant microorganisms are frequently poorly characterized (either metabolically or physiologically) under uncontrolled environmental conditions (e.g. when grown with rich undefined media).

Several reported data show that large variation of fluxes in metabolism can be achieved through a change in growth conditions, even at a constant growth rate. The *in vivo* flux changes are, however, only partially reflected in changes of enzyme levels. Only large changes lead to differences in enzyme levels, but these differences are much smaller than the variations in the fluxes (Vriezen and van Dijken, 1998). Even though fluxes differed by a factor of 45, the maximum difference found in the enzyme levels was only a factor of 3. Both the largest flux differences and the largest variations in enzyme levels were observed for the glycolytic enzymes. On the bases of: *(i)* the higher level of all glycolytic enzymes detected in low oxygen chemostat cultures, and *(ii)* the reduction only in the levels of hexokinase and pyruvate kinase, in low glucose chemostat cultures, Vriezen and van Dijken (1998) conclude that the down-regulation of glycolysis is only partially accomplished at the level of enzyme synthesis and flux modulation is primarily effected via concentrations of substrates, activators, and inhibitors (bottom layer according to Fig. 1.6). Similar conclusions were reached by Sierkstra *et al.* (1992) who measured mRNA levels and activities of glycolytic enzymes (HK, PFK, PGI, PGM), glucose-6-phosphate dehydrogenase, and glucose-regulated enzymes (pyruvate decarboxylase, pyruvate dehydrogenase, invertase, alcohol dehydrogenase), in glucose-limited continuous cultures of an industrial strain of *S. cerevisiae* at different dilution rates. The analysis showed that there is no clear correlation between enzyme activity and mRNA levels despite the PGI1 mRNA fluctuated and a slight decrease in PGI was registered at increasing growth rates. Despite the fact that increasing the dilution rate led to an increase in glycolytic flux, the activity of most enzymes remained constant. According to these results, Sierkstra *et al.* (1992) suggested that the glycolytic flux is not regulated at either transcriptional or translational levels under the conditions studied, but through effectors of enzymes (i.e. allosteric or protein phosphorylation mechanisms, Fig. 1.6, bottom layer).

During mixed-substrate cultivation of *S. cerevisiae*, the observed differential regulation of enzyme activities indicated that glucose is preferentially used as the starting material for biosynthesis and ethanol production preferentially as a dissimilatory substrate for energy production (and a source of acetyl-CoA for biosynthesis). Only when the ATP requirement for glucose assimilation was completely met by oxidation of ethanol, were the enzymes required for assimilation of ethanol into compounds with more than two carbon atoms synthesized (De Jong-Gubbels *et al.*, 1995). In contrast to the coordinated expression pattern that was observed for the key enzymes of ethanol assimilation, activities in cell-free extracts of the glycolytic enzymes, phosphofructokinase and pyruvate kinase, exhibited little variation with the ethanol to glucose ratio. This is consistent with the view that regulation of these enzyme activities is largelly controlled by the concentrations of substrates and products and/or by allosteric enzyme modification (De Jong-Gubbels *et al.*, 1995).

Thus, a major area of research in future must determine whether the transcriptional level of regulation operates effectively under defined environmental conditions. As well as this, its articulation with the regulatory network of metabolism (e.g. allosteric or covalent modification) remains a major area of study. Unsolved problems include the following: *(i)* under which conditions do the transcriptional regulatory networks exert control? *(ii)* for which metabolic blocks or specific reactions? *(iii)* how main biological processes (division and differentiation) influence these regulatory circuits? *(iv)* are they always the same, or *(v)* do they change following cell division or differentiation? All these are open questions which remain to be elucidated (see Chapter 7).

Expression levels of more than 6200 yeast genes using high-density oligonucleotide arrays for monitoring the expression of total mRNA populations, has been performed (Wodicka *et al.*, 1997). More than 87% of all yeast mRNAs were detected in *S. cerevisiae* cells grown in rich medium. The expression comparison between cells grown in rich and minimal media identified a relatively small number of genes with dramatically different expression levels. Many of the most highly expressed genes are common to cells grown under both conditions, including genes encoding well-known "house-keeping" enzymes (e.g. *PGK1, TDH3, ENO2, FBA1,* and *PDC1*), structural proteins such as actin, and many ribosomal proteins (Wodicka *et al.*, 1997). However, many of the most highly expressed genes and those with the largest differences, under the conditions explored by Wodicka *et al.* (1997), are of unknown function. Another large- scale screen of genes that express differentially during the life cycle of

S. cerevisiae at different subcellular locations, has been performed in diploid strains containing random *lacZ* insertions throughout the genome (Burns *et al.*, 1994). Powerful as they are for knowing which genes are expressed under defined environmental conditions, these approaches do not give information about the effectiveness of the functioning of the products of those genes in cells, e.g. actual reaction rates, enzyme kinetics, activation, inhibition (Fig. 1.6).

Another timely topic for MCE, concerns the relationship between the different cellular processes, although these are frequently treated separately. Although fragmentary, some experimental evidence exists that shows a link, e.g. between cell division, cell differentiation, and catabolite repression in yeast (Mónaco *et al.*, 1995; Aon and Cortassa, 1995, 1997, 1998; Cortassa and Aon, 1998; Cortassa *et al.*, 2000). We will deal with this subject more thoroughly later (see Chapters 4 and 7).

Transduction and Intracellular Signalling

Cells exhibit many distinct signalling pathways that allow them to react to environmental stimuli. Overall, experimental evidence increasingly favors the idea that metabolism occurs strongly associated to the dynamic cellular scaffolds, and that these insoluble matrices and their dynamics are in turn influenced strongly by the dynamics of chemical reactions (see Aon *et al.*, 2000, for a review). At this point, transduction and coherence merge since the cytoskeleton fulfils all the requirements for systems to self-organize, and as a prevailing and ubiquitous (macro)molecular cytoplasmic network, may function as a link between, e.g. the stress-sensing and the stress-transduction mechanisms.

Intracellular signalling is orchestrated through a large number of components by way of their interactions and their spatial relationships (Weng *et al.*, 1999). Networking and nonlinearity of the input-output transfer characteristics results in several emergent properties that the individual pathways do not have. In this way supplying an additional store of information within the intracellular biochemical reactions of signalling pathways becomes possible. Although mutations or altered gene expression can result in persistent activation of protein kinases, connections between pre-existing pathways may also result in persistently activated protein kinases capable of eliciting biological effects. Based on the considerations mentioned above the following properties arise: *(i)* extended signal duration; *(ii)* activation of feedback loops that confers on the system the ability to regulate output for considerable periods; this is achieved by allowing coupling between fast and slow responses; *(iii)* definition of threshold stimulation for biological effects, since signals of defined amplitude and duration are required to evoke a physiological response; *(iv)* multiple signal outputs that provide a filter mechanism to ensure that only appropriate signals are translated into alterations in biological behavior (Bhalla and Iyengar, 1999).

In systems where two signalling pathways interact through a feedback loop, the amplitude and duration of the extracellular signal may determine the sustained activation of the system (Bhalla and Iyengar, 1999; Weng *et al.*, 1999). The system may behave as a self-organized bistable that defines a threshold level of stimulation, provided an autocatalytic loop (the nonlinearity), exists in the network (Aon and Cortassa, 1997). These bistable systems may be also deactivated, thus the emergent properties of this feedback system define not only the amplitude and duration of the extracellular signal required to be activated but the magnitude and duration need to be deactivated as well (Bhalla and Iyengar, 1999). Robustness then is one of the emergent features of these networks because of their ability to deliver, once activated, a constant output in a manner unaffected by small fluctuations caused by activating or deactivating events.

Self-organized Emergent Phenomena

Even in integrated form, the genomic databases will never allow us to go from DNA sequence to function directly; this arises as a consequence of the extremely dynamic and self-organized nature of cells and organisms. As already stressed major cellular functions and activities are self-organized to such an extent as to relegate the importance of genome-based information (see Lloyd, 1992, 1998; and Aon and Cortassa, 1997, for reviews).

In the second half of the twentieth century, most of the fundamental details of the spatio-temporal organization of living systems has become established. For the first time, we have at hand a biophysical theory with which to approach the quantitative and qualitative analysis of the organized complexity that characterize living systems. Two main foundations of this biophysical theory are self-organization (Nicolis and Prigogine, 1977; Haken, 1978; Kauffman, 1989, 1995), and Dynamic Systems Theory (Rosen, 1970; Abraham, 1987). Self-organization is deeply rooted in non-equilibrium thermodynamics (Nicolis and Prigogine, 1977), and the kinetics of nonlinear systems, whereas Dynamic Systems Theory derives from the geometric theory of dynamical systems created by Poincaré (Abraham and Shaw, 1987).

By applying this biophysical theory of biological organization to successively more complicated systems (i.e. artificial, artificial-biological-oriented, or biological, Lloyd, 1992), it became clear that self-organization is a fundamental and necessary property of living systems. Conditions under which self-organization appears (Aon and Cortassa, 1997) are:

1. openness to fluxes of energy and matter;
2. the operation of coupled processes through some common intermediate;
3. the occurrence of at least one process that exhibits a kinetic nonlinearity.

Kinetically, biological systems in general, and cells in particular, are nonlinear because of multiple interactions between their components, e.g. protein-protein, feedbacks (e.g. substrate) inhibition, feedforward (product) activation, cross-activation or cross-inhibition. In the case of chemical reactions, they may be arranged in linear, branched, cyclic pathways or as combinations of these basic configurations. Within this framework, metabolic pathways with different topologies (linear, circular) and sources of nonlinear kinetics (allostery, stoichiometric autocatalysis) were compared as energy converters for thermodynamic efficiency (Fig. 1.7) (see also Aon and Cortassa, 1997). The converters were investigated for possible energetic advantage under oscillatory dynamics as compared with asymptotic steady state behavior as the rate of substrate uptake was systematically varied. Essentially, oscillatory dynamics arising from stoichiometric autocatalysis allowed an advantage in thermodynamic efficiency, whereas there was not such an advantage for oscillations where the underlying nonlinear mechanism was allostery despite the same pathway topology (i.e. linear). Furthermore, efficiency decreased in the case of a cyclic topology with a similar nonlinear mechanism as the source of the oscillations. These experiments demonstrate that the topology of the metabolic pathway also has a pivotal influence on the putative energetic advantage exerted by periodic dynamics (Cortassa *et al.*, 1990; Aon and Cortassa, 1991, 1997). The higher thermodynamic efficiency recorded in the linear pathway with a stoichiometric autocatalytic mechanism is due to a phase shift between the simultaneous maxima in catabolic and anabolic fluxes and a minimum in the chemical potential of substrate conversion (Fig. 1.7b).

Thus due to the intrinsic built-in nonlinearities in the network of chemical reactions, and their compartmentation, biological systems are able to change their dynamic behavior, i.e. to bifurcate toward new steady states, or attractors. The example shown in Fig. 1.7 illustrates the fact that biological systems are able to display diverse mechanisms of energetic adaptation when challenged under different environmental conditions.

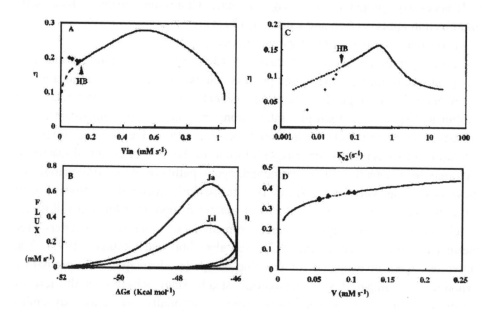

Figure 1.7. Thermodynamic performance of metabolic pathways with different topologies and feedback mechanisms as a function of the rate of substrate uptake by the cell. The putative advantage of oscillatory dynamics on thermodynamic performance was investigated on three different metabolic pathways: (A-B) A stoichiometric model of glycolysis in which the nonlinear mechanism is the autocatalytic feedback exerted on glucose phosphorylation by the stoichiometry of the ATP production of the anaerobic functioning of the glycolytic pathway. The glycolytic pathway is also an example of a *linear-branched topology. (A)* The stability analysis as a function of the substrate input V_{in} provided the steady state concentrations of the metabolites from which the thermodynamic functions could be calculated. *(B)* Phase space relations of the fluxes (J_a, ATP synthesis; J_{sl}, glucose degradation) with the input force (ΔG_s, the chemical potential of glucose conversion) in a limit cycle, which is included to explain the reversion of the declining tendency of thermodynamic efficiency in (a), as glycolysis approaches a Hopf bifurcation (HB). The parameters correspond to those described in (a) for the lowest V_{in} values represented by diamonds. *(C)* A model of the phosphotransferase (PTS) system in bacteria that represents a global *circular topology* including a covalent cycle and branches. The parameter k_{e2} represents the rate at which the phosphate group of the covalent cycle A-A~P is transferred to glucose when transported into the cell (Aon and Cortassa, 1997). Thermodynamic efficiency was analyzed as a function of k_{e2}. The total nucleotide (C_A) and phosphate pool (P_T) were both 10 mM, whereas the total sum of substrate A involved in the covalent modification cycle (TA) was 1 mM. *(D)* An allosteric model described in Goldbeter (1996). Thermodynamic efficiency was studied as a function of V, the rate of substrate input.

For selected points (diamonds in A, C, D) in the oscillatory domain (dashed lines) the simulation of the temporal evolution of the system enabled the computation of the thermodynamic efficiency, η, in order to examine the energetic advantage of oscillatory dynamics in all models represented in A, C, D (see eqs. in Aon and Cortassa, 1997). In all figures, the plein and dashed lines represent stable and unstable steady states, respectively. (Reproduced from *Dynamic Biological Organization*, 1997, pp. 310–311, Ch. 8, Fig. 8.11, Aon and Cortassa, Chapman & Hall, London, with kind permission from Kluwer Academic Publishers).

These common features exhibited by different cellular systems, as well as their basic dynamic nature, support the notion that not all the information on the spatio-temporal organization in cells and organisms is encoded in the DNA. Furthermore, the catalytic role of cytoskeletal polymers beyond their well-known structural one (Clegg, 1984, 1991; Aon and Cortassa, 1995, 1997; Aon *et al.*, 2000), and the crowded nature of the intracellular environment (Gomez-Casati *et al.*, 1999, 2000; Aon *et al.*, 2000), add new feasible behavioral possibilities. Among them, the appearance of self-organized behavior in coupled dynamic subsystems, e.g. microtubule polymerization dynamics and coupled-enzymatic reactions (Cortassa *et al.*, 1994; Aon *et al.*, 1996), or the triggering of ultrasensitive catalytic behavior in a crowded milieu (Gomez-Casati *et al.*, 1999, 2000; Aon *et al.*, 2001).

In summary, cellular systems are able to show coherent as well as emergent properties (Lloyd, 1998; see Aon and Cortassa, 1997, for a review). On this basis, for example, enzymatic activities can show threshold effects (e.g. bistability or multiple stationary states) during entrainment by another autocatalytic process (e.g. cytoskeleton polymerization). Indeed, the observation that the dynamics of enzymatic reactions in turn may be entrained by the intrinsic dynamic instability of cytoskeleton polymers (Mitchison and Kirschner, 1984) may throw light on the physiological role of so-called "structural elements" (Cortassa and Aon, 1996; Aon and Cortassa, 1997).

The combinatorial binding of transcription factors to other proteins as well as to high and low affinity DNA sites, or when the spacing between DNA binding sites is altered (Miklos and Rubin, 1996), points out another example. Thus, large responses may result from small changes in the concentrations of transcriptional components, or phosphorylation of transcriptional factors (Aon and Cortassa, 1997; Gomez-Casati *et al.*, 1999; Aon *et al.*, 2001).

Knowledge of all the putative regulatory mechanisms and the intrinsic nonlinear nature of biological processes should be taken into account when designing a modified organism for a specific biotransformation. In this respect, mathematical simulation of the desired process by attempting to incorporate into the model formulation as many variables as necessary to account for the known dynamical behavior will enhance the predictive power of the analysis and enlighten the decisions.

Homeodynamics and Coherence

Biological systems, e.g. cells, operate under *homeodynamic* conditions (see Chapter 5). These refer to the continuous motion of the system's dynamics, and its propensity for shifting between attractors at bifurcation points, based on its intrinsic dynamic properties (Fig. 1.4) (Lloyd *et al.*, 2001). The latter is given by the general tendency of the system to self-organize and the characteristic nature of its nonlinear kinetic mechanisms as well as the extensive degree of coupling between processes.

Models based on the stoichiometry of metabolic pathways for flux calculations apply for stable (asymptotic) steady states, and currently do not account for regulatory mechanisms (e.g. product inhibition, effector activation). Thus, the dynamic behavior of metabolic networks cannot be assessed by these type of models. In Chapters 4 and 5 we introduce some basic concepts and tools for dealing with the dynamics of biological processes.

Under homeodynamic conditions, living systems spatio-temporally coordinate their functioning by essentially *top-bottom* or *bottom-up* mechanisms. The former are represented by circadian and ultradian rhythms with clock characteristics whereas the latter emerge from the intrinsic, autonomous dynamics of the integrated mass-energy-information-carrying networks that represent living systems (Fig. 1.6). Organized complexity is generated by the cross-talk between these two opposing, but complementary, flows of information. For instance circadian rhythms may arise from the intrinsic dynamics of mass-energy-information-carrying networks; the latter being in turn influenced by those rhythms through, e.g. entrainment.

Relevant to the functional behavior of cells or tissues, are the mechanisms through which dynamic organization under homeodynamic conditions is achieved; MCE of these dynamic aspects has not yet begun (see Chapter 5).

Coherence may arise from the synchronization in space and time of molecules, or the architecture of supramolecular or supracellular structures, through self-organization, in an apparent, "purposeful", functional way. Under coherent behavioral conditions, spatially distant, say cytoplasmic regions, function coordinately, spanning spatial coordinates larger than the molecular or supramolecular realms, and temporal relaxation times slower than those usually considered typical of molecular or supramolecular components (Aon *et al.*, 2000; Lloyd *et al.*, 2001). Thus a major functional consequence is brought about: the qualitative behavior of the system changes through the scaling of its spatio-temporal coordinates (Aon and Cortassa, 1997).

It has been proposed that the ultradian clock has timing functions that provide a time base for intracellular coordination (Lloyd, 1992). In the latter sense, ultradian oscillations are a potentially coherence-inducer of the top-bottom type.

We previously stated that dynamically organized phenomena are visualized as being spatio-temporal coherent. Several mechanisms are at the origin of this bottom-up coherence, that we briefly describe. Waves of, e.g. second messengers or ions, may arise through a combination of amplification in biochemical reaction networks and its spatial spreading through diffusion or percolation. Amplification may arise at instabilities in the dynamics of biochemical reactions given by autocatalysis through allosteric or ultrasensitive mechanisms (Aon *et al.*, 2000, 2001; Gomez-Casati *et al.*, 1999, 2000). Under these conditions, transduction (sensitivity amplification) and coherence (spatio-temporal waves) mutually cooperate.

Dynamic supramolecular organization of cytoskeleton components gives rise to sophisticated spatial organization and intricate fractal geometries in cells. The cellular cytoskeleton fulfils all the requirements for systems to self-organize, i.e. it is open to fluxes of matter (proteins) and energy (GTP), and nonlinearity is provided by autocatalysis during polymerization or by the so-called "dynamic instability" that implies the catastrophic depolymerization of microtubules (Kirschner and Mitchinson, 1986). We have offered an interpretation of microtubular dynamic instability as an example of a bistable irreversible transition. The dynamic coupling between changes in the geometry of cytoskeleton organization and of enzymatic reactions taking place concomitantly, produces entrainment of one system by the other, in a global bistable switch (Cortassa and Aon, 1996; Aon *et al.*, 1996).

Figure 1.8 shows that the degree of cytoskeleton polymerization may exert systemic effects on the glycolytic flux. Stability and bifurcation analyzes were performed on a mathematical model that couples the dynamics of assembly-disassembly of microtubular protein (MTP) to the glycolytic pathway and the branch to the tricarboxylic acid cycle, ethanolic fermentation and the pentose phosphate pathway.

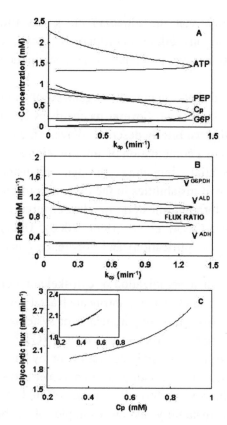

Figure 1.8. Stability and bifurcation analyzes of a mathematical model that couples the cycle of assembly-disassembly of MTP and a metabolic network, as a function of the depolymerization, k_{dp}, and polymerization, k_{pol}, constants of MTP. A mathematical model was formulated, comprising eleven ordinary differential equations; eight of them representing metabolites concentration whereas the other three describe the concentration of polymerized or non-polymerized (GTP-bound) MTP and the oligomeric status of pyruvate kinase (Lloyd *et al.*, 2001; Aon and Cortassa, 2001, unpublished results). The steady state behavior and stability of the mathematical model as well as the existence of bifurcation points were analyzed as a function of parameters related with the assembly-disassembly of MTP. The main panels show the results obtained through variation of k_{dp} whereas the insets correspond to the simulations performed with k_{pol}. Shown are the steady state behavior of metabolites concentration as state variables of the model (A) and the fluxes through individual enzymatic steps as functions of state variables (B). For the sake of clarity only some metabolites and fluxes are represented, placed at different levels of the metabolic network. Also depicted are the phase plane analyzes of the glycolytic flux as a function of the level of polymerized MTP, Cp (C). The inset of panel C belongs to the results obtained with model simulations as a function of k_{pol} and the axis corresponds to the same variables and units as in main panels.

The level of polymerized MTP was changed through variation of parameters related with MTP dynamics, i.e. the rate constant of microtubules depolymerization, k_{dp}, or polymerization, k_{pol} (Fig. 1.8). On the one hand, the enzymatic rates and metabolites concentration are entrained by the MTP polymeric status (Fig. 1A, B) whereas phase plane analysis shows that the flux through glycolysis coherently changes with the level of polymerized MTP (Fig. 1C) as reflected by augmenting or diminishing the glycolytic flux concomitant with an increase or decrease of polymer levels. Further studies showed that high levels of the polymer decrease the negative control exerted by the branch to the PP pathway over glycolysis, thus allowing higher fluxes through the network (Lloyd *et al.*, 2001; Aon and Cortassa, 2001, unpublished).

Biochemically and thermodynamically, cellular metabolism may be represented as a set of catabolic and anabolic fluxes coupled to each other through energy-transducing events. In this framework, pathway stoichiometry constitutes a built-in source of nonlinear behavior able to give rise to homeodynamic behavior, either monotonic or periodic. Thus, metabolic flux coordination is an expression of the dynamic cellular organization. Flux coordination was shown to be involved in the regulation of cell growth, division and sporulation in *S. cerevisiae*. Thus subcellular structural remodeling occurs through the degree of coupling between carbon and energy fluxes (Aon and Cortassa, 1995, 1997; Cortassa *et al.*, 2000; Aon, J.C. and Cortassa, 2001).

The ability of biological systems, either unicellular or multicellular, to exhibit rhythmic behavior in the ultradian domain is a fundamental property because of its potential role as an inducer of coherence, e.g. entrainment and coordination of intracellular functions (Lloyd, 1992). In systems exhibiting chaos, many possible motions are simultaneously present. In fact, since a chaotic system traces a strange attractor in the phase space, in principle a great number of unstable limit cycles are embedded in this attractor, each characterized by a distinct number of oscillations per period (Fig. 1.4) (Peng *et al.*, 1991; Shinbrot *et al.*, 1993; Aon *et al.*, 2000). For *homeodynamic* operation, biological systems exhibiting chaotic dynamics need only small perturbations of their parameters to selecting stable periodic outputs (Lloyd and Lloyd, 1995), in this way facilitating the system's dynamic motion between alternative states or operational modes.

Chapter 2

Matter and Energy Balances

Mass Balance

Bioprocess optimization through manipulation of metabolism, requires a quantitative analysis of metabolic flux performance under defined conditions. As an integrated network of chemical reactions, metabolism obeys the general laws of chemistry, e.g. mass and energy conservation.

Material balance relies on the first law of thermodynamics. All the mass entering a system must be recovered, transformed or unchanged, at the output of a system, or at the end of a process. There are essentially two approaches to mass and energy balances. The first approach is to perform a global balance, that takes into account the overall input and output of the system, whilst disregarding the nature of the transformations taking place inside the living organism participating in the bioprocess. This approach is called **Black Box,** and is the most commonly used by process engineers to evaluate the performance of a biotransformation. A second approach, let us call it "**Light Grey**", considers most reactions taking part in the metabolic machinery of an organism that participates in a particular bioprocess, and calculate energy and material balances thereof. This chapter will apply the concepts of mass balance from the level of the whole process taking place in the bioreactor ("black box") to the "light grey" description. of microbial growth associated with the metabolic changes occurring within growing cells (see Chapter 1 section: *Black and grey boxes: levels of description of metabolic behavior in microorganisms*).

From the point of view of the exchange of matter and energy, most bioprocesses operating either at laboratory or industrial scales, may be classified as closed or open. A bioprocess that operates under **batch** conditions, corresponds to a closed system that does not exchange matter except gases (usually O_2 and CO_2). Another operational mode is the **fed-batch** fermenter that is again open to gas exchange, but allows only the input of matter (e.g. substrates, vitamins). Completely open systems are those continuous processes consisting of input of matter and energy and exit of biomass, products, and energy. Substrates in excess with respect to the limiting one, also flow out the system. In this sense, cells are non-equilibrium systems that depend on the rates at which energy or matter, or both, are fed into the system (Aon and Cortassa, 1997). To support steady state operation, cells must dissipate energy and be continuously fed with matter.

The operation costs of continuous processes, are much higher than batch or fed-batch ones. In each case, an evaluation of the yield is required to make a decision about the type of operating mode that should be applied. As an example, streptomycin production by *Streptomyces griseus* or antibodies from hybridoma cells, are currently run in batch cultures. Ethanol fermentations from sugar cane or beet molasses are carried out in fed-batch cultures, whereas continuous cultures are run for acetic acid production from ethanol by *Acetobacter aceti* (Doran, 1995).

A cell participating in a bioprocess behaves as an open system because of the continuous uptake of substrates and excretion of metabolic products. Even under the condition of "endogenous" metabolism (when internal stores are consumed because of exhaustion of nutrients available in the environment), the cellular system excretes products. The substrates and products in the environment of cellular systems either may be freely changing or "clamped" depending on the experimental conditions. For instance, in batch culture the medium is continuously changing, whereas in a chemostat culture operating at steady state, both external and internal metabolite concentrations are fixed. This latter condition facilitates the rigorous application of some quantitative techniques germane to the purpose of MCE, i.e. MFA and MCA (see Chapter 4).

General Formulation of Mass Balance

The first law of thermodynamics states that during any chemical or physical change (subatomic reactions are excluded), there is no modification of the total amount of matter participating. Expressed in formal terms:

$$M_{input} + M_{prod} - M_{output} - M_{consum} = M_{accumul} \tag{2.1}$$

where M_{input} stands for the input of matter to the system; M_{prod} is the matter produced during the process; M_{output} stands for the output of matter from the system; M_{consum} is the matter consumed; $M_{accumul}$ is the matter accumulated in the system during the process.

Equation 2.1 is of general validity, irrespective of the type of system or process under study. It can be applied to chemical species in a fermentor, the mass of a molecular species, or to elemental balances such as those for carbon. It may also be applied to processes occurring exclusively inside cells, e.g. synthesis of cytoplasmic proteins or those undergoing processes that comprise intra- and extra-cellular stages, as is the case for sucrose transformation into ethanol by yeast alcoholic fermentation. The latter involves the cleavage of sucrose by secreted invertase, followed by its catabolism through cytoplasmic glycolysis and end-product excretion.

There are particular conditions under which simplifications of Eq. 2.1 can be considered.

(*i*) When there are no reactions involved, but only physical changes, i.e. there is no matter being produced or consumed, then M_{prod} and M_{consum} cancel out, and Eq. 2.1 can be reduced to:

$$M_{input} - M_{output} = M_{accumul} \tag{2.2}$$

This condition is without interest for the processes under consideration in this book, and will not be further analyzed.

(*ii*) Under steady state conditions, in which the concentration of chemicals and the other system variables are constant, the amount of accumulated matter, $M_{accumul}$, is zero. The mass balance expression can then be written as follows:

$$M_{input} + M_{prod} - M_{output} - M_{consum} = 0 \tag{2.3}$$

Equation 2.3 is also valid for the total mass in a closed system, as well as for the elemental balance of nitrogen or sulphur, in batch processes with chemoorganotroph organisms; it is not valid for carbon since CO_2 is a gaseous compound and leaves the system.

Integral and Differential Mass Balances

Different methods are used to perform a mass balance, depending on the operational mode of the process being considered: (*i*) in batch and fed-batch processes; the total and/or initial input of matter feeding the system is readily available, and a so-called "integral balance" can be performed; (*ii*) in continuous processes, where the fluxes of matter being exchanged by the system are calculated, a "differential mass balance" is then performed. This can be formalized as follows:

$$\begin{bmatrix} matter \\ input\ to \\ the\ system \end{bmatrix} + \begin{bmatrix} matter \\ generated\ in \\ the\ system \end{bmatrix} - \begin{bmatrix} matter \\ output\ from \\ the\ system \end{bmatrix} - \begin{bmatrix} matter \\ consumed\ in \\ the\ system \end{bmatrix} = \begin{bmatrix} matter \\ accumulated \\ in\ the\ system \end{bmatrix} \tag{2.4}$$

that in the steady state takes the form:

$$
\begin{bmatrix} matter \\ input\ to \\ the\ system \end{bmatrix} + \begin{bmatrix} matter \\ generated\ in \\ the\ system \end{bmatrix} = \begin{bmatrix} matter \\ output\ from \\ the\ system \end{bmatrix} - \begin{bmatrix} matter \\ consumed\ in \\ the\ system \end{bmatrix} \qquad (2.5)
$$

A note of caution should be raised at this point. While performing a mass balance (as well as other types of calculations), care should be taken to keep the units consistent. Analysis of units is very helpful for the detection of sources of error or missing information.

As our focus is on Metabolic and Cellular Engineering, we will leave here the general formulation of material balance to focus on processes involving metabolic transformations in cells. For a deeper explanation on the general matter of bioprocess balancing, the reader is referred to the book of Doran (1995) for useful examples.

Growth Stoichiometry and Product Formation

As any process operating in closed or open systems, the growth of a cell or its division to generate another cell, may be considered from the point of view of mass balances. From this viewpoint, the two approaches defined under Section: *Material and energy balances,* will be developed. On the one hand, the "**black box**" approach, which just looks at the overall inputs and outputs of cells without taking into account the biochemical transformations occurring inside them. The overall mass balance equation for growth of a chemoorganotroph organism without product formation (Fig. 2.1), may be expressed as follows:

$$
C_w H_x O_y N_z + a\ O_2 + b\ H_g O_h N_i \Rightarrow c\ CH_\alpha O_\beta N_\delta + d\ CO_2 + e\ H_2O \qquad (2.6)
$$

where a, b, c, d and e are the stoichiometric coefficients of the overall biomass synthesis reaction, and $\alpha, \beta, \delta, g, h, i, w, x, y$ and z, the elemental formula coefficients. Figure 2.1 illustrates the **black box** approach for biosynthesis of heterotrophic, phototrophic, or chemolithotrophic microbial mass (Fig. 2.1, panels A, B and C, respectively). The carbon and nitrogen substrates, the energy source, and the electron acceptor, used by the microorganism (O_2 or an organic compound) are taken into account. It also shows the by-products associated with biomass production, e.g. CO_2 and H_2O, as in the case of chemoorganotrophs. Such by-products differ for the microbial processes in phototrophic organisms, performing photosynthesis, and thereby O_2 instead of CO_2 is evolved. On the other hand, chemolitotrophic organisms use reduced inorganic compounds, e.g. SH_2 as electron donor and energy source (Fig. 2.1).

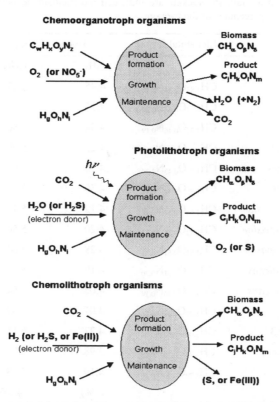

Figure 2.1. Black box scheme of growth and product formation exhibited by different kind of organisms. In the scheme are emphasized the nature of the carbon source (i.e. organic compound in top panel, or inorganic, CO_2, in middle and bottom panels), the nitrogen source ($H_gO_hN_i$), the electron donor (in top panel the C-source, H_2O or H_2, SH_2 or Fe(II) in middle and bottom panels) and the products of metabolism. Chemoorganotrophs use O_2 as electron acceptor, alternatively they may use NO_3^- whereas chemolithotrophs and photolithotrophs use endogenous organic compounds.

Other elements are used to produce microbial biomass, e.g. P, S, Na, K, Ca, and trace elements (Fe, Cu, Mn), but the uptake of these elements may be considered negligible in mass balance calculations. When performing elemental analysis, a small fraction of ash is usually formed. This fraction corresponds mainly to P and S residues, although other elements may also account for ash.

Table 2.1. Elemental composition and generalized degree of reduction of biomass with respect to various nitrogen sources. The degree of reduction was calculated as indicated by Eq. 2.16. The differences between ammonia, nitrate and N_2 arise because of their γ values of -3, $+5$ and 0, respectively are taken as null. In brackets are indicated the standard deviation of the degree of reduction expressed as percentage of the average.

Organism	Elemental formula	Degree of reduction (γ)		
		NH_3	HNO_3	N_2
Candida utilis	$CH_{1.83}O_{0.54}N_{0.10}$	4.45	5.25	4.75
Candida utilis	$CH_{1.87}O_{0.56}N_{0.20}$	4.15	5.75	4.75
Candida utilis	$CH_{1.83}O_{0.46}N_{0.19}$	4.34	5.86	4.91
Klebsiella aerogenes	$CH_{1.75}O_{0.43}N_{0.22}$	4.23	5.99	4.89
Klebsiella aerogenes	$CH_{1.73}O_{0.43}N_{0.24}$	4.15	6.07	4.87
Klebsiella aerogenes	$CH_{1.75}O_{0.47}N_{0.17}$	4.30	5.66	4.81
Saccharomyces cerevisiae	$CH_{1.64}O_{0.52}N_{0.16}$	4.12	5.40	4.60
Saccharomyces cerevisiae	$CH_{1.83}O_{0.56}N_{0.17}$	4.20	5.56	4.71
Saccharomyces cerevisiae	$CH_{1.81}O_{0.51}N_{0.17}$	4.28	5.64	4.79
Paracoccus denitrificans	$CH_{1.81}O_{0.51}N_{0.20}$	4.19	5.79	4.79
Paracoccus denitrificans	$CH_{1.51}O_{0.46}N_{0.19}$	4.02	5.54	4.59
Escherichia coli	$CH_{1.77}O_{0.49}N_{0.24}$	4.07	5.99	4.79
Pseudomonas C12B	$CH_{2.00}O_{0.52}N_{0.23}$	4.27	6.11	4.96
Aerobacter aerogenes	$CH_{1.83}O_{0.55}N_{0.25}$	3.98	5.98	4.73
Average	$CH_{1.79}O_{0.50}N_{0.20}$	4.19	5.78	4.79
		(3%)	(4.5%)	(2.1%)

Reprinted from Roels, J.A., *Energetics and Kinetics in Biotechnology*, pp. 30–41. ©copyright 1983 Elsevier Science, with permission of the author.

Roels (1983) has collated the compositions of a number of microorganisms, and derived values for a mean "chemical formula" (Table 2.1). It can be appreciated that a "chemical formula" is not a distinctive feature of an organism as different organisms may have the same "formula". Moreover, different elemental formula coefficients may occur within a species when it grows under different environmental conditions, e.g. with substrates of different degree of reduction, under aerobiosis or anaerobiosis, or with different nitrogen sources.

As an example, the chemical formula of *Pseudomonas mendocina* growing under microaerophilic conditions was calculated from its elemental composition (C, 43.5%; H, 6.5%; N, 11.5%; O, 30.3%) as follows (Verdoni *et al.*, 1992):

$$C: \frac{43.5}{12} = 3.6 \quad ; \quad H: \frac{6.5}{1} = 6.5 \quad ; \quad N: \frac{11.5}{14} = 0.82 \quad ; \quad O: \frac{30.3}{16} = 1.89$$

The formula $C_{3.6}H_{6.5}O_{1.89}N_{0.82}$ can be recalculated as that of a molecule containing a single atom carbon: $CH_{1.81}O_{0.53}N_{0.23}$ thereby corresponds to a molecular weight of 25.5.

To find out the stoichiometric coefficients of biomass synthesis, once the chemical formula is known, an "elemental balance" must be performed. In such an elemental balance only the main biomass components are considered, i.e. C, N, O and H:

C balance $\qquad\qquad w = c + d$ $\qquad\qquad\qquad\qquad$ (2.7a)

N balance $\qquad\qquad z + b\,i = c\,\delta$ $\qquad\qquad\qquad\qquad$ (2.7b)

O balance $\quad y + 2\,a + b\,h = c\,\beta + 2\,d + e$ $\qquad\qquad$ (2.7c)

H balance $\qquad\quad x + b\,g = c\,\alpha + 2\,e$ $\qquad\qquad\qquad$ (2.7d)

In order to calculate the stoichiometric coefficients, the system of Eqs. 2.7 cannot be determined because five coefficients are unknown, and only four equations are available. A fifth equation is provided by the respiratory quotient, RQ (Eq. 2.8), that in physiological terms tells us about the type of metabolism displayed by the organism, i.e. respiratory or respiro-fermentative.

$$RQ = \frac{moles\ of\ CO_2\ produced}{moles\ of\ O_2\ consumed} = \frac{d}{a} \qquad\qquad (2.8)$$

Respiratory quotients range from 1.0 for oxidative breakdown of carbohydrates to larger values for fermentative metabolism, and in ethanolic fermentation can attain values of 10. However, the minimal value depends on the degree of reduction of the substrate being oxidized. For instance, in methane or ethanol it falls to values of 0.5 and 0.67, respectively.

In practice, the stoichiometric coefficient e (Eqs. 2.7.c and 2.7.d) is usually very difficult to measure, due to the large excess of water in the aqueous systems where most bioprocesses are studied. A change in water concentration is subjected to a large experimental error, because it is negligible in comparison with the water present in the medium. A way to circumvent this problem is to perform a balance of the reducing power of available electrons (see below).

If growth is associated with other organic products, as happens under anaerobic conditions (when the electron acceptor is an oxidized metabolic intermediate), then:

$$C_wH_xO_yN_z + b\,H_gO_hN_i \Rightarrow c\,CH_\alpha O_\beta N_\delta + d\,CO_2 + e\,H_2O + f\,C_jH_kO_lN_m \qquad (2.9)$$

where $C_jH_kO_lN_m$ stands for a by-product (e.g. ethanol, lactic acid or a secondary metabolite).

The formation of product adds a right-hand term to the system of Eqs. 2.7; this additional term contains an unknown coefficient, f.

The balance equation for biomass and product formation for photosynthetic organisms (e.g. higher plants, green algae and photosynthetic bacteria, e.g. cyanobacteria) reads as follows:

$$a\,CO_2 + b\,H_gO_hN_i + c\,H_2O \Rightarrow d\,CH_\alpha O_\beta N_\delta + e\,O_2 + f\,C_jH_kO_lN_m \qquad (2.10)$$

While for the chemolitotrophic organisms, such as sulphur iron or hydrogen bacteria (Fig. 2.1 C) is:

$$a\,CO_2 + b\,H_gO_hN_i + c\,SH_2 \Rightarrow d\,CH_\alpha O_\beta N_\delta + e\,S + f\,C_jH_kO_lN_m \qquad (2.11)$$

Biomass and Product Yields

There are some coefficients that, in spite of being dependent on the physiological status of the growing organism, may be assessed from a complete "black box" approach. We are referring here to biomass and product yields with respect to the consumed carbon substrate.

The biomass yield is defined as the amount of biomass synthesized per mole or gram of carbon substrate. In formal terms and expressed on a mass basis:

$$Y_{XC} = \frac{grams\ of\ produced\ cells}{grams\ of\ consumed\ substrate} = \frac{c\,(MW\ cells)}{(MW\ substrate)} \qquad (2.12)$$

If product is also formed:

$$Y_{PC} = \frac{grams\ of\ synthesized\ product}{grams\ of\ consumed\ substrate} = \frac{f\,(MW\ product)}{(MW\ substrate)} \qquad (2.13)$$

Y_{XC} and Y_{PC} are, in fact, phenomenological coefficients. The mechanistic stoichiometric coefficients (deduced in, e.g. a "light grey" approach) may be higher, depending upon culture conditions and the degree of coupling between anabolism and catabolism. In fact, part of the consumed substrate is directed toward biomass synthesis, and part is devoted to energy generation to drive biosynthetic reactions. Thus, growth yields are meaningful parameters that can also be derived from the biochemistry of microbial growth (see Chapter 4).

The maximal yield of product obtained from a given substrate, may be predicted in the black box approach from an electron balance (see below). In physiological experiments, the yield may be obtained from the slope of the specific rate of substrate consumption as a function of the growth rate, as indicated by the following expression:

$$\frac{dS}{dt} = m_S + \frac{\mu}{Y_{XS}} \qquad (2.14)$$

the left hand term being the instantaneous rate of substrate consumption, and in the right hand term, m_S, the maintenance coefficient, μ, the growht rate, and Y_{XS}, the yield of biomass on substrate S. We shall not go into details of Eq. 2.14 here; a more detailed treatment of the relationship between substrate consumption, growth, and yield is addressed in Chapter 3.

Electron Balance

In metabolic networks a large number of reactions imply changes in the redox state of the compounds involved. In fact, catabolism includes either complete or partial oxidation of reduced compounds, coupled to the synthesis of high-energy-transfer potential intermediates, e.g. ATP. Thus, the electron balance is at the interphase between mass and energy balances.

Reducing power is another quantity subjected to conservation laws. In fact, what is actually conserved is the number of "available" electrons; i.e. those able to be transferred to oxygen during the combustion of a substance to CO_2, H_2O, and a nitrogenous compound (Doran, 1995). For many organic compounds this is an exergonic process (i.e. ΔG is negative, see below Section: *Energy balance*) as exemplified in Table 2.2 (Roels, 1983). It is convenient to calculate the available electrons (valence) from the elemental balance: C (+4), O (−2), H (+1). In the case of nitrogen, its valence depends on the reference compound, whether it is NH_3 (valence -3 for N), NO_3^- (+5 for N), or molecular nitrogen N_2 (0 for N).

The degree of reduction of an organic compound is defined as the mole numbers of available electrons contained in the amount corresponding to one gram-mole C of the compound (Table 2.2). This is a very useful measurement since it tells us whether a substrate can act as electron donor (and thereby supply energy for biomass growth) with a negative thermodynamic efficiency, i.e. without requiring catabolism to operate in order to fulfil the energy requirement of anabolism. If a compound is more reduced than biomass (degree of reduction is *ca.* 4 with NH_3 as the reference nitrogen compound), a negative thermodynamic efficiency of biomass growth will result (Table 2.2, see also Section: *An energetic view of microbial metabolism*). The degree of reduction is calculated from the addition of the product of the coefficients of each element in the chemical formula times the valence of that element, as follows:

$$\gamma_S = 4w + x - 2y - 3z \quad \text{for a substrate} \tag{2.15}$$

and

$$\gamma_B = 4 + \alpha - 2\beta - 3\delta \quad \text{for biomass} \tag{2.16}$$

Ammonia is considered here as the N source. The degree of reduction of CO_2, or H_2O, is zero. The electron balance is a useful criterion for the identification of the kind of by-products that can be excreted to the extra-cellular medium when the carbon balance of a given culture is investigated.

Theoretical Oxygen Demand

The oxygenation of a bioprocess is an important condition for cell growth. Under oxygen limitation, bioprocesses operate under anaerobic or microaerophilic conditions and thereby the type of metabolism displayed by the organism.

Additionally, the stoichiometry of the energy coupling of the respiratory chain may vary according to the oxygen tension level maintained under particular growth conditions.

The P:O ratio, namely the mole numbers of ATP synthesized per mole of electron pairs transported across the respiratory electron transport chain, is difficult to assess experimentally since several different possibilities of electron transport stoichiometries exist (see below). For instance, in chemostat cultures of *P. mendocina* at very low oxygen tensions, a P:O ratio of 3 had to be considered for closing energy balance calculations (Verdoni *et al.*, 1992).

Table 2.2 Heat of combustion and free energy of combustion at standard conditions (unit molality, 298°K, 1 atmosphere) and at a pH 7. Free enthalpy of combustion to gaseous CO_2, liquid water and N_2 at unit molality of all reactants. The values in the sixth and seventh columns were obtained from the fourth and fifth columns, respectively, divided by the degree of reduction and the number of C atoms in the molecular formula.

Compound	Formula	Degree of reduction, γ	ΔG^{o}_{c}	ΔH^{o}_{c} (kJ mole^{-1})	$\Delta G^{o}_{c}/\gamma$	$\Delta H^{o}_{c}/\gamma$
Formic acid	CH_2O_2	2	281	255	140.5	127.5
Acetic acid	$C_2H_4O_2$	4	894	876	111.8	109.5
Propionic acid	$C_3H_6O_2$	4.67	1533	1529	109.4	109.2
Butyric acid	$C_4H_8O_2$	5	2173	2194	108.7	109.7
Valeric acid	$C_5H_{10}O_2$	5.2	2813	2841	108.2	109.3
Palmitic acid	$C_{16}H_{32}O_2$	5.75	9800	9989	106.5	108.6
Lactic acid	$C_3H_6O_3$	4	1377	1369	114.8	114.1
Gluconic acid	$C_6H_{12}O_7$	3.67	2661		121	
Pyruvic acid	$C_3H_4O_3$	3.33	1140		114	
Oxalic acid	$C_2H_2O_4$	1	327	246	163.5	123
Succinic acid	$C_4H_6O_4$	3.5	1599	1493	114.2	106.7
Fumaric acid	$C_4H_4O_4$	3	1448	1337	120.7	111.3
Malic acid	$C_4H_6O_5$	3	1444	1329	120.3	110.8
Citric acid	$C_6H_8O_7$	3	2147	1963	119.3	109.1
Glucose	$C_6H_{12}O_6$	4	2872	2807	119.7	117
Methane	CH_4	8	818	892	102.3	111.5
Ethane	C_2H_6	7	1467	1562	104.8	111.6
Propane	C_3H_8	6.67	2108	2223	105.4	111.1
Pentane	C_5H_{12}	6.4	3385	3533	105.8	110.4
Ethene	C_2H_4	6	1331	1413	110.9	117.8
Ethyne	C_2H_2	5	1235	1301	123.5	130.1
Methanol	CH_4O	6	693	728	115.5	121.3
Ethanol	C_2H_6O	6	1319	1369	109.9	114.1
iso-Propanol	C_3H_8O	6	1946	1989	108.1	110.5
n-Butanol	$C_4H_{10}O$	6	2592	2680	108	111.7
Ethylene glycol	$C_2H_6O_2$	5	1170	1181	117	118.1
Glycerol	$C_3H_8O_3$	4.67	1643	1663	117.4	118.7
Glucitol	$C_6H_{14}O_6$	4.33	3084	3049	118.6	117.3

Table 2.2 (*Continued*)

Compound	Formula	Degree of reduction, γ	ΔG^o_c	ΔH^o_c (kJ mole^{-1})	$\Delta G^o_c/\gamma$	$\Delta H^o_c/\gamma$
Acetone	C_3H_6O	5.33	1734	1793	108.4	112.1
Formaldehyde	CH_2O	4	501	572	125.3	142.9
Acetaldehyde	C_2H_4O	5	1123	1168	112.3	116.8
Butyraldehyde	C_4H_8O	5.5	2407		109.4	
Alanine	$C_3H_7NO_2$	5	1642	1707	109.5	113.8
Arginine	$C_6H_{14}N_4O_2$	5.67	3786	3744	111.3	110
Asparagine	$C_4H_8N_2O_3$	4.5	1999	1936	111.1	107.6
Glutamic acid	$C_5H_9NO_4$	4.2	2315	2250	110.2	107.1
Aspartic acid	$C_4H_7NO_4$	3.75	1686	1608	112.4	107.2
Glutamine	$C_5H_{10}N_2O_3$	4.8	2628	2570	109.5	107.1
Glycine	$C_2H_5NO_2$	4.5	1011	974	112.3	108.2
Leucine	$C_6H_{13}NO_2$	5.5	3565	3588	108	108.7
Isoleucine	$C_6H_{13}NO_2$	5.5	3564	3588	108	108.7
Lysine	$C_6H_{14}N_2O_2$	5.67		3684		108.3
Histidine	$C_6H_9N_3O_2$	4.83		3426		118.2
Phenylalanine	$C_9H_{11}NO_2$	4.78	4647	4653	108	108.2
Proline	$C_5H_9NO_2$	5		2735		109.4
Serine	$C_3H_7NO_3$	4.33	1502	1455	115.6	112.0
Threonine	$C_4H_9NO_3$	4.75	2130	2104	112.1	110.7
Tryptophane	$C_{11}H_{12}N_2O_2$	4.73	5649	5632	108.6	108.3
Tyrosine	$C_9H_{11}NO_3$	4.56	4483	4437	109.2	108.1
Valine	$C_5H_{11}NO_2$	5.4	2920	2920	108.2	108.2
Thymine	$C_5H_6N_2O_2$	4.4		2360		107.3
Adenine	$C_5H_5N_5$	5		2780		111.2
Guanine	$C_5H_5N_5O$	4.6	2612	2500	113.6	108.7
Cytosine	$C_4H_5N_3O$	4.75		1828		96.2
Uracil	$C_4H_4N_2O_2$	4		1688		105.5
Hydrogen	H_2	2	238	286	119	143
Graphite	C	4	394	394	98.5	98.5
Carbon monoxide	CO	2	257	283	128.5	141.5
Ammonia	NH_3	3	329	383	109.7	127.7
Ammonium ion	NH_4^+		356	383		
	NO	-2	86.6	22	-43.3	-11

Table 2.2 (*Continued*)

Compound	Formula	Degree of reduction, γ	ΔG^o_c	ΔH^o_c (kJ mole^{-1})	$\Delta G^o_c/\gamma$	$\Delta H^o_c/\gamma$
Nitrous acid	HNO_2	-3	81.4		-27.1	
Nitric acid	HNO_3	-5	7.3	-30	-1.5	6
Hydrazine	N_2H_4	4	602.4	622	150.6	155.5
Hydrogen sulfide	H_2S	2	323	247	161.5	123.5
Sulfurous acid	H_2SO_3	-4	-249.4	-329	62.4	82.3
Sulfuric acid	H_2SO_4	-6	-507.4	-602	84.6	100.3
Biomass	$CH_{1.8}O_{0.5}N_{0.2}$	4.8	541.2	560	112.8	116.7

Reprinted from Roels, J.A. *Energetics and Kinetics in Biotechnology*, pp. 30–41. ©copyright 1983, Elsevier Science, with permission of the author.

An oxygen growth yield has also been defined as oxygen is also a substrate, mainly a catabolic one, acting as the final electron acceptor of the respiratory chain. Y_{O2} is a physiological parameter containing implicit information about the effciency of oxidative phosphorylation (Stouthamer and van Verseveld, 1985).

If the stoichiometric coefficients of Eqs. 2.6 and 2.9 are known, the conservation law of available electrons can be written as:

$$w\,\gamma_S - 4\,a = c\,\gamma_B + j\,\gamma_P f \qquad (2.17)$$

where a depicts the stoichiometric coefficient for oxygen (the oxygen demand for the growth reaction) whether associated or not with product formation. In Eq. 2.17, apart from CO_2 and O_2; the degree of reduction of NH_3 is also taken as zero. In that case, the degree of reduction of biomass and product should be computed with respect to NH_3 as reference compound.

The electrons provided by the substrate acting as electron donor may be transferred to oxygen, biomass, and the product. The distribution of the electrons transferred may be calculated from the following expression (Doran, 1995):

$$1 = \frac{4\,a}{w\,\gamma_S} + \frac{c\,\gamma_B}{w\,\gamma_S} + \frac{f\,j\,\gamma_P}{w\,\gamma_S} \qquad (2.18)$$

Table 2.3. Anabolic reactions in step I from fermentable and gluconeogenic carbon sources The overall reactions leading to the synthesis of key intermediary metabolites from glucose, glycerol, lactate, pyruvate, acetate and ethanol are depicted with emphasis in ATP, reducing power, inorganic phosphate and CO_2 participation.

Glucose + ATP	$\Rightarrow$ Hexose-6-P + ADP
Glucose + ATP + 2 $NADP^+$	$\Rightarrow$ Ribose-5-P + ADP + CO_2 + 2 NADPH
Glucose + ATP + 4 $NADP^+$	$\Rightarrow$ Erythrose-4-P + ADP + 2 CO_2 + 4 NADPH
Glucose + 2 ATP	$\Rightarrow$ 2 Triose-P + 2 ADP
Glucose + 2 NAD^+	$\Rightarrow$ 2 3-P-glycerate + 2 NADH
Glucose + 2 NAD^+ + 2 Pi	$\Rightarrow$ 2 P-Enolpyruvate + 2 NADH
Glucose + 2 ADP + 2 NAD^+ + 2 Pi	$\Rightarrow$ 2 Pyruvate + 2 ATP + 2 NADH
Glucose + 2 ADP + 4 NAD^+ + 2 CoA	$\Rightarrow$ 2 Acetyl CoA + 2 ATP + 4 NADH + 2 CO_2
Glucose + 2 CO_2 + 2 NAD^+	$\Rightarrow$ 2 Oxalacetate + 2 NADH
Glucose + ADP + 3 NAD^+ + $NADP^+$	$\Rightarrow$ α-ketoglutarate + 3 NADH + NADPH + ATP + CO_2
2 Glycerol + 2 ATP + 2 NAD^+	$\Rightarrow$ Hexose-6-P + 2 ADP + 2 NADH
2 Glycerol + 2 ATP + 2 NAD^+ + 2 $NADP^+$	$\Rightarrow$ Ribose-5-P + 2 ADP + CO_2 + 2 NADPH + 2 NADH
2 Glycerol + 2 ATP + 2 NAD^+ + 4 $NADP^+$	$\Rightarrow$ Erythrose-4-P + 2 ADP + 2 CO_2 + 4 NADPH + 2 NADH
Glycerol + ATP + NAD^+	$\Rightarrow$ Triose-P + ADP + NADH
Glycerol + 2 NAD^+ + Pi	$\Rightarrow$ 3-P glycerate + 2 NADH
Glycerol + 2 NAD^+ + Pi	$\Rightarrow$ 2 P-Enolpyruvate + 2 NADH
Glycerol + ADP + 2 NAD^+ + Pi	$\Rightarrow$ Pyruvate + ATP + 2 NADH
Glycerol + ADP + 3 NAD^+ + CoA + Pi	$\Rightarrow$ Acetyl CoA + ATP + 3 NADH + CO_2
Glycerol + CO_2 + 2 NAD^+	$\Rightarrow$ Oxalacetate + 2 NADH
2 Glycerol + ADP + 5 NAD^+ + $NADP^+$	$\Rightarrow$ α-ketoglutarate + 5 NADH + NADPH + ATP + CO_2
2 Pyruvate + 6 ATP + 2 NADH	$\Rightarrow$ Hexose-6-P + 6 ADP + 2 NAD^+ + 5 Pi
5 Pyruvate + 15 ATP + 5 NADH	$\Rightarrow$ 3 Ribose-5-P + 15 ADP + 12 Pi + 5 NAD^+
4 Pyruvate + 12 ATP + 4 NADH	$\Rightarrow$ 3 Erythrose-4-P + 12 ADP + 9 Pi + 4 NAD^+
Pyruvate + 3 ATP + NADH	$\Rightarrow$ Triose-P + 3 ADP + NAD^+ + 2 Pi
Pyruvate + 2 ATP	$\Rightarrow$ 3-P glycerate + 2 ADP + Pi
Pyruvate + 2 ATP	$\Rightarrow$ 2 P-Enolpyruvate + 2 ADP + Pi

Table 2.3 (*Continued*)

Pyruvate + NAD$^+$ + CoA	$\Rightarrow$	Acetyl CoA + NADH + CO$_2$
Pyruvate + CO$_2$ + ATP	$\Rightarrow$	Oxalacetate + ADP + Pi
2 Pyruvate + ATP + NAD$^+$ + NADP$^+$	$\Rightarrow$	α-ketoglutarate + NADH + NADPH + ADP + Pi + CO$_2$
2 Lactate + 6 ATP + 2 NADH + 4 Fe(III) cyt b$_2$	$\Rightarrow$	Hexose-6-P + 6 ADP + 2 NAD$^+$ + 5 Pi + 4 Fe(II) cyt b$_2$
5 Lactate + 15 ATP + 5 NADH + 10 Fe(III) cyt b$_2$	$\Rightarrow$	3 Ribose-5-P + 15 ADP + 12 Pi + 5 NAD$^+$ + 10 Fe(II) cyt b$_2$
4 Lactate + 12 ATP + 4 NADH + 8 Fe(III) cyt b$_2$	$\Rightarrow$	3 Erythrose-4-P + 12 ADP + 9 Pi + 4 NAD$^+$ + 8 Fe(II) cyt b$_2$
Lactate + 3 ATP + NADH + 2 Fe(III) cyt b$_2$	$\Rightarrow$	Triose-P + 3 ADP + NAD$^+$ + 2 Pi + 2 Fe(II) cyt b$_2$
Lactate + 2 ATP + 2 Fe(III) cyt b$_2$	$\Rightarrow$	3-P glycerate + 2 ADP + Pi + 2 Fe(II) cyt b$_2$
Lactate + 2 ATP + 2 Fe(III) cyt b$_2$	$\Rightarrow$	2 P-Enolpyruvate + 2 ADP + Pi + 2 Fe(II) cyt b$_2$
Lactate + 2 Fe(III) cyt b$_2$	$\Rightarrow$	Pyruvate + 2 Fe(II) cyt b$_2$
Lactate + NAD$^+$ + CoA + 2 Fe(III) cyt b$_2$	$\Rightarrow$	Acetyl CoA + NADH + CO$_2$ + 2 Fe(II) cyt b$_2$
Lactate + CO$_2$ + ATP + 2 Fe(III) cyt b$_2$	$\Rightarrow$	Oxalacetate + ADP + Pi + 2 Fe(II) cyt b$_2$
2 Lactate + ATP + NAD$^+$ + NADP$^+$ + 4 Fe(III) cyt b$_2$	$\Rightarrow$	α-ketoglutarate + NADH + NADPH + ADP + Pi + CO$_2$ + 4 Fe(II) cyt b$_2$
4 Acetate + 8 ATP + 4 NAD$^+$ + 2 FAD$^+$	$\Rightarrow$	Hexose-6-P + 4 ADP + 4 AMP + 4 NADH + 2 FADH + 11 Pi + CO$_2$
10 Acetate + 20 ATP + 10 NAD$^+$ + 5 FAD$^+$	$\Rightarrow$	3 Ribose-5-P + 10 ADP + 10 AMP + 27 Pi + 10 NAD$^+$ + 5 FADH + 5 CO$_2$
8 Acetate + 16 ATP + 8 NAD$^+$ + 4 FAD$^+$	$\Rightarrow$	3 Erythrose-4-P + 8 ADP + 8 AMP + 21 Pi + 8 NAD$^+$ + 4 FADH + 4 CO$_2$
2 Acetate + 4 ATP + 2 NAD$^+$ + FAD$^+$	$\Rightarrow$	Triose-P + 2 ADP + 2 AMP + 5 Pi + 2 NADH + FADH + CO$_2$
2 Acetate + 3 ATP + 2 NAD$^+$ + FAD$^+$	$\Rightarrow$	3-P glycerate + 2 AMP + 4 Pi + ADP + 2 NADH + FADH + CO$_2$
2 Acetate + 3 ATP + 2 NAD$^+$ + FAD$^+$	$\Rightarrow$	P-enolpyruvate + 2 AMP + 4 Pi + ADP + 2 NADH + FADH + CO$_2$
2 Acetate + 2 ATP + NADP$^+$ + 2 NAD$^+$ + FAD$^+$	$\Rightarrow$	Pyruvate + 2 AMP + 4 Pi + NADPH + 2 NADH + FADH + CO$_2$
Acetate + ATP + CoA	$\Rightarrow$	Acetyl CoA + AMP + 2 Pi
2 Acetate + 2 ATP + 2 NAD$^+$ + FAD$^+$	$\Rightarrow$	Oxalacetate + 2 AMP + 4 Pi + 2 NADH + FADH
3 Acetate + 3 ATP + 2 NAD$^+$ + NADP$^+$ + FAD$^+$	$\Rightarrow$	α-ketoglutarate + 2 NADH + NADPH + 3 AMP + 6 Pi + CO$_2$

Table 2.3 (*Continued*)

4 Ethanol + 8 ATP + 8 NAD$^+$ + 4 NADP$^+$ + 2 FAD$^+$	$\Rightarrow$	Hexose-6-P + 4 ADP + 4 AMP + 8 NADH + 4 NADPH + 2 FADH + 11 Pi + CO$_2$
10 Ethanol + 20 ATP + 20 NAD$^+$ + 10 NADP$^+$ + '5 FAD$^+$	$\Rightarrow$	3 Ribose-5-P + 10 ADP + 10 AMP + 27 Pi + 20 NADH + 10 NADPH + 5 FADH + 5 CO$_2$
8 Ethanol + 16 ATP + 16 NAD$^+$ + 8 NADP$^+$ + 4 FAD$^+$	$\Rightarrow$	3 Erythrose-4-P + 8 ADP + 8 AMP + 21 Pi + 16 NADH + 8 NADPH + 4 FADH + 4 CO$_2$
2 Ethanol + 4 ATP + 4 NAD$^+$ + 2 NADP$^+$ + FAD$^+$	$\Rightarrow$	Triose-P + 2 ADP + 2 AMP + 5 Pi + 4 NADH $^+$ + 2 NADPH + FADH + CO$_2$
2 Ethanol + 3 ATP + 4 NAD$^+$ + 2 NADP$^+$ + FAD$^+$	$\Rightarrow$	3-P glycerate + 2 AMP + 4 Pi + ADP + 4 NADH + 2 NADPH + FADH + CO$_2$
2 Ethanol + 3 ATP + 4 NAD$^+$ + 2 NADP$^+$ + FAD$^+$	$\Rightarrow$	P-enolpyruvate + 2 AMP + 4 Pi + ADP + 4 NADH + 2 NADPH + FADH + CO$_2$
2 Ethanol + 2 ATP+ 3 NADP$^+$ + 4 NAD$^+$ + FAD$^+$	$\Rightarrow$	Pyruvate + 2 AMP + 4 Pi + 3 NADPH + 4 NADH + FADH + CO$_2$
Ethanol + ATP + NADP$^+$ + NAD$^+$ + CoA	$\Rightarrow$	Acetyl CoA + AMP + 2 Pi + NADPH + NADH
2 Ethanol + 2 ATP + 4 NAD$^+$ + 2 NADP$^+$ + FAD$^+$	$\Rightarrow$	Oxalacetate + 2 AMP + 4 Pi + 4 NADH + 2 NADPH + FADH
3 Ethanol + 3 ATP + 6 NAD$^+$ + 4 NADP$^+$ + FAD$^+$	$\Rightarrow$	α-ketoglutarate + 6 NADH + 4 NADPH + 3 AMP + 6 Pi + CO$_2$

Assuming that all electrons were transferred to biomass, the maximal amount that could be formed from a substrate, may be calculated as follows:

$$c_{max} = \frac{w \, \gamma_S}{\gamma_B} \qquad (2.19)$$

The c_{max} can be never attained; as in all chemoorganotrophic organisms a part of the carbon substrate has to be oxidized to fuel the biological system with ATP to drive biosynthesis; thus, not all their electrons can be transfered to biomass.

Likewise, if no biomass is synthesized, the maximal amount of product that could be formed is given by:

$$f_{max} = \frac{w \, \gamma_S}{j \, \gamma_P} \qquad (2.20)$$

Equation 2.20 should not be used alone, but together with the stoichiometric elemental balance given by Eqs. 2.7. Aberrant results may otherwise be obtained, such as a value of c_{max} of almost 2 for methane (limited in fact to 1 because of its C content, Erickson *et al.*, 1978).

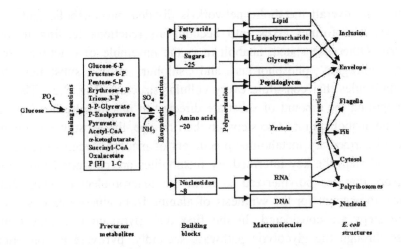

Figure 2.2. Overall scheme of metabolism in leading to the synthesis of *Escherichia coli* from glucose as C source. The figure emphasized four levels of "reactions" according to their function in the organization of a bacterial cell: (*i*) The fuelling reactions are those producing the key precursors metabolites (Step 1 of anabolism) as well as those supplying ATP and reducing power for biosynthesis. The latter involves catabolic and amplibolic pathways. (*ii*) Biosynthesis reactions refer to those pathways leading from the key intermediary metabolites to the monomers precursors of macromolecules, e.g. amino acids, nucleotides, sugars, fatty acids. These pathways constitute what is called the Step 2 of anabolism. (*iii*) Polymerisation reactions refer to the synthesis of the polymeric chains of proteins, nucleic acids, polysaccharides and lipids. (*iv*) Assembly reactions that involve the modification, transport of macromolecules to specific locations in the cell and their association to form cellular structures. The size of the boxes is proportional to their participation in the biomass. (Reproduced from Ingraham, Maaloe and Neidhardt, 1983, with © permission from Sinauer Associates, Inc.).

Opening the "Black Box". Mass Balance as the Basis of Metabolic Flux Analysis

Anabolic fluxes

Metabolic Flux Analysis (MFA) is an approach fundamental to MCE. MFA is based on mass balance and stoichiometry of metabolic pathways (Vallino and Stephanopoulos, 1993; Varma and Palsson, 1994; Cortassa *et al.*, 1995; Bonarius *et al.*, 1996).

In this section, we would like to introduce the basis of a method for estimating metabolic fluxes, in order to illustrate another application of material balances. This introduction will allow us to develop later the rationale followed for anabolic and catabolic flux calculations.

First, the overall metabolic network is divided into anabolic and catabolic pathways. The former takes into account all reactions leading to biomass synthesis. Otherwise stated, anabolism is the ensemble of reactions providing carbon, nitrogen, phosphorous and sulphur, from substrates to the macromolecules that constitute every cellular structure (Fig. 2.2). The strategy for computing the amount of substrate directed to macromolecules is to further divide the anabolism into two steps; in the first one, the substrates are converted into key intermediary metabolites precursors of macromolecules (Table 2.3). In the second step, the key intermediary metabolites are converted into monomer precursors that are polymerized to form the macromolecules (Fig. 2.2). For instance, to account for the synthesis of alanine from glucose as C source, two lumped steps are considered. In the first one, pyruvate is synthesized from glucose through the glycolytic pathway; secondly, pyruvate is converted into alanine through a specific pathway. The reader is invited to consult a biochemistry textbook for the details of reactions and metabolic pathways involved.

Although central metabolic pathways may show peculiar differences in some organisms, their major features are shared by nearly all of them. Table 2.3 summarizes the reactions leading from carbon substrates commonly used by *S. cerevisiae* to each of the key intermediates: Hexose-6-P, Triose-3-P, Ribose-5-P, Erythrose-4-P, 3-P-glycerate, P-enolpyruvate, Pyruvate, Oxalacetate, α-ketoglutarate, Acetyl Co-A. Together with the stoichiometry for the conversion of a C source into each key intermediate, the relative amounts of ATP, NAD(H) or NADP(H), CO_2 and Pi are balanced. The formation of Ribose-5-P and Erythrose-4-P is concomitant with the reduction of NADP, since the oxidative branch of the pentose phosphate pathway is involved when glucose or glycerol act as C source. With ethanol, a large amount of NADPH is also synthesized together with all key intermediates because of the cofactor requirement of acetaldehyde dehydrogenase. The demand of phosphorylation energy required to fuel the anabolic Step I is rather less with glucose or glycerol than on the other C sources, a comparison made on the bases of a similar amount of key intermediates obtained. It should be noted that from the elemental composition of biomass, only carbon is balanced. Table 2.4 shows the second step of anabolism that transforms key intermediates into the monomer precursors of macromolecules (aminoacids, nucleotides, monosaccharides and fatty acids).

Nitrogen, sulphur and phosphorous are involved in the reactions of the second step (Step II of anabolism, Table 2.4). Oxygen and hydrogen are not balanced, as H_2O or H^+ were not included in the equations of Table 2.4 due to the fact that they are either very difficult to measure (H_2O) and highly dependent on pH.

The pathways in Step II are exclusively anabolic ones, whereas the reactions of Step I are shared between anabolism and catabolism; the pathways to which they belong are thereby called amphibolic pathways.

To compute the amount of substrate, or of each of the intermediates required to build up a gram of biomass, both the relative amounts of macromolecules and their monomeric composition, should be known. The required amount of a monomer precursor of macromolecules is related, through the corresponding stoichiometric coefficient, to the amount of the key precursor intermediate (Table 2.5). Likewise, the amount of a key intermediary metabolite is related to the C source through the corresponding stoichiometric coefficient (Table 2.6). Tables 2.5 and 2.6 constitute the basis for the construction of stoichiometric matrices in MFA (see Chapter 4).

In *S. cerevisiae,* all pathways in Step II are the same, irrespective of the C-source, with the exception of the synthesis of serine and glycine from acetate (Cortassa *et al.,* 1995). Different phosphorylation and redox potentials are necessary for the synthesis of monomer precursors. For instance, amino acids synthesis requires both ATP and reducing equivalents as NADPH, whereas in nucleotide synthesis ATP demand is much larger, and in lipid (fatty acid) synthesis NADPH dependence is greater.

Table 2.7 presents some selected examples of the amount of intermediates, and/or C substrate, needed to synthesize an amino acid (arginine), a nucleotide (UTP), and a monosaccharide (glucose); these are a direct function of the growth rate. In chemostat cultures, the latter is equal to the dilution rate, D, when the culture is at steady state (see Chapters 3-5).

Table 2.4 Anabolic reactions in Step II from each key intermediary metabolite. Anabolic reactions leading from the key intermediary metabolites to the synthesis of monomers precursor of macromolecules, e.g. amino acids, nucleotides, fatty acids and sugarsare depicted with emphasis in ATP, reducing power, ammonia, sulphate, inorganic phosphate and CO_2 participation.

α-ketoglutarate + NADPH + NH_4^+	$\Rightarrow$ Glutamate + $NADP^+$
α-ketoglutarate + NADPH +ATP + 2 NH_4^+	$\Rightarrow$ Glutamine + ADP + Pi + $NADP^+$
α-ketoglutarate + 3 NADPH + ATP + NH_4^+	$\Rightarrow$ Proline + ADP + Pi + 3 $NADP^+$
α-ketoglutarate + CO_2 + 4 NADPH + 5 ATP + 4 NH_4^+	$\Rightarrow$ Arginine + 4 ADP + AMP + 4 Pi + PPi + 4 $NADP^+$
α-ketoglutarate + AcCoA + NAD^+ + 3 NADPH + ATP + 2 NH_4^+	$\Rightarrow$ Lysine + CO_2 + AMP + PPi + 3 $NADP^+$ + NADH
Oxalacetate + NADPH + NH_4^+	$\Rightarrow$ Aspartate + $NADP^+$
Oxalacetate + 2 ATP + NADPH + 2 NH_4^+	$\Rightarrow$ Asparagine + 2 ADP + 2 Pi + $NADP^+$
Oxalacetate + 2 ATP + 3 NADPH + NH_4^+	$\Rightarrow$ Threonine + 2 ADP + 2 Pi + 3 $NADP^+$
Oxalacetate + Formate + 4 ATP + NADH + 7 NADPH + $SO4^=$ + NH_4^+	$\Rightarrow$ Methionine + AMP + 3 Pi + PPi + 3 ADP + 7 $NADP^+$ + NAD^+
Pyruvate + NADPH + NH_4^+	$\Rightarrow$ Alanine + $NADP^+$
2 Pyruvate + 2 NADPH + NH_4^+	$\Rightarrow$ Valine + 2 $NADP^+$ + CO_2
Pyruvate + Oxalacetate +2 ATP+ 5 NADPH + NH_4^+	$\Rightarrow$ Isoleucine + 5 $NADP^+$ + CO_2 + 2 ADP + 2 Pi
2 Pyruvate + AcCoA+ 2 NADPH + NAD^+ + NH_4^+	$\Rightarrow$ Leucine + 2 $NADP^+$ + NADH + 2 CO_2
3-P-glycerate+ NAD^+ + NADPH + NH_4^+	$\Rightarrow$ Serine + $NADP^+$ + NADH + Pi
3-P-glycerate + NAD^+ + NH_4^+ + ADP	$\Rightarrow$ Glycine + NADH + ATP + Formate
3-P-glycerate+ NAD^+ + 5 NADPH + $SO4^=$ + NH_4^+	$\Rightarrow$ Cysteine + 5 $NADP^+$ + NADH + Pi
2 PEP + Ery4P + ATP + 2 NADPH + NH_4^+	$\Rightarrow$ Phenylalanine + 4 Pi + 2 $NADP^+$ + CO_2 + ADP
2 PEP + Ery4P + ATP + NAD^+ + 2 NADPH + NH_4^+	$\Rightarrow$ Tyrosine + 4 Pi $^+$ + NADH + 2 $NADP^+$ + CO_2 + ADP
2 PEP + Ery4P + R5P + 3 ATP + NAD^+ + 2 NADPH + 2 NH_4^+	$\Rightarrow$ Tryptophane + 4 Pi + NADH + 2 $NADP^+$ + CO_2 + 2 ADP + AMP + PPi + Pyruvate

Table 2.4 (*Continued*)

R5P + 5 ATP + 3 NH_4^++ 2 NADPH + 2 NAD^+ + Formate	$\Rightarrow$	Histidine + 2 AMP + 3 ADP + 2 PPi + 2 NADH + 2 $NADP^+$ + 3 Pi
Ribose-5-P+ 3-P-glycerate + 9 ATP+ NAD^+ + 5 NH_4^++ 2 NADPH + CO_2 + Formate	$\Rightarrow$	AMP + AMP + 8ADP + PPi + 9 Pi + 2 $NADP^+$ + NADH
Ribose-5-P+ 3-P-glycerate + 10 ATP + 2 NAD^+ + 5 NH_4^+ + NADPH + CO_2 + Formate	$\Rightarrow$	GMP + 2 AMP + 9 ADP + 2 PPi + 10 Pi + $NADP^+$ + 2 NADH
Ribose-5-P + Oxalacetate + NAD^+ + 4 ATP + 2 NH_4^++ NADPH	$\Rightarrow$	UMP + AMP + 3 ADP + NADH + 3 Pi + PPi
Ribose-5-P +Oxalacetate + NAD^+ +8 ATP + 3 NH_4^++ NADPH	$\Rightarrow$	CTP + AMP + 7 ADP + NADH + 7 Pi + PPi
Glucose 6 P + ATP + $Glucan_{n-1}$	$\Rightarrow$	$Glucan_n$ + ADP + PPi
Glucose 6 P + ATP + $Manan_{n-1}$	$\Rightarrow$	$Manan_n$ + ADP + PPi
Glucose 6 P + ATP + $Glycogen_{n-1}$	$\Rightarrow$	$Glycogen_n$ + ADP + PPi
6 AcCoA + 5 ATP + 5 NADPH	$\Rightarrow$	Lauric acid(12) + 5 ADP + 5 $NADP^+$ + 5 Pi
8 AcCoA + 7 ATP + 8 NADPH	$\Rightarrow$	Palmitoleic acid(16:1) + 7 ADP + 8 $NADP^+$ + 7 Pi
9 AcCoA + 8 ATP + 9 NADPH	$\Rightarrow$	Oleic acid(18:1) + 8 ADP + 9 $NADP^+$ + 8 Pi
Triose-P + NADH	$\Rightarrow$	Glycerol-3-P
Glycerol-3-P + Fatty acid + ATP	$\Rightarrow$	Acyl-glycerol + Pi + AMP + PPi

Table 2.5 Stoichiometries of production of key intermediary metabolites from different carbon sources

Carbon substrate or Intermediate	Cost of making 1 μmol of each key intermediary metabolite[a] (μmol/ μmol)									
	G6P	R5P	E4P	TP	3PG	PEP	PY	AcCoA	OAA	α KG
Glucose	1[b]	1	1	0.5	0.5	0.5	0.5	0.5	0.5	1
NADH	0	0	0	0	-1[b]	-1	-1	-2	-1	-3
NADPH	0	-2	-4	0	0	0	0	0	0	-1
ATP	1	1	1	1	0	0	-1	-1	0	-1
CO2	0	-1	-2	0	0	0	0	-1	1	-1
Glycerol	2	2	2	1	1	1	1	1	1	2
NADH	-2	-2	-2	-1	-2	-2	-2	-3	-2	-5
NADPH	0	-2	-4	0	0	0	0	0	0	-1
ATP	2	2	2	1	0	0	-1	-1	0	-1
CO2	0	-1	-2	0	0	0	0	-1	1	-1
Pyruvate	2	1.67	1.33	1	1	1	1	1	1	2
NADH	2	1.67	1.33	1	0	0	0	-1	0	-1
NADPH	0	0	0	0	0	0	0	0	0	-1
ATP	6	5	4	3	2	2	0	0	1	1
CO2	0	0	0	0	0	0	0	-1	1	-1
Lactate	2	1.67	1.33	1	1	1	1	1	1	2
NADH	2	1.67	1.33	1	0	0	0	-1	0	-1
NADPH	0	0	0	0	0	0	0	0	0	-1
ATP	6	5	4	3	2	2	0	0	1	1
CO2	0	0	0	0	0	0	0	-1	1	-1
Ethanol	4	3.33	2.67	2	2	2	2	1	2	3
NADH	-4	-3.33	-2.67	-2	-3	-3	-2	-1	-3	-4
NADPH	-4	-3.33	-2.67	-2	-2	-2	-3	-1	-2	-4
FADH	-2	-1.67	-1.33	-1	-1	-1	-1	0	-1	-1
ATP	12	10	8	6	5	5	4	2	4	6
CO2	-2	-1.67	-1.33	-1	-1	-1	-1	0	0	-1
Acetate	4	3.33	2.67	2	2	2	2	1	2	3
NADH	0	-1.67	-1.33	0	-1	-1	0	0	-1	-1
NADPH	0	0	0	0	0	0	-1	0	0	-1
FADH	-2	-1.67	-1.33	-1	-1	-1	-1	0	-1	-1
ATP	12	10	8	6	5	5	4	2	4	6
CO2	-2	-1.67	-1.33	-1	-1	-1	-1	0	0	-1

[a] Carbon, ATP, NAD(P)H, FADH and CO_2 demand for production of one mol of each key intermediary metabolite is computed according to the balance of intermediates through the pathways indicated in Table 2.3.
[b] positive (negative) numbers indicate consumption (production) of the intermediate together with the key intermediary metabolite.

(Fluxes of carbon, phosphorylation, and redox intermediates during growth of *Saccharomyces cerevisiae* on different carbon sources, Cortassa, Aon and Aon, *Biotech. Bioengin*. 47, 193–208. ©copyright 1995 John Wiley & Sons, Inc. Reprinted by permission of Wiley-Liss, Inc., a subsidiary of John Wiley & Sons, Inc.).

Table 2.6. Monomer composition of each macromolecular fraction of yeast cells

Monomer	Relative amounts present in *S. cerevisiae*[a]		Cost of making 1 µmol of each of the macromolecular monomers (µmol/ µmol)					
	(mol)	Metabolite[b]	NADPH	NADH	~P[c]	NH$_3$	CO$_2$	SO$_4$
Amino acid								
alanine	1.000	1 PY	1	0	0	1	0	0
arginine	0.351	1 αKG	4	-1	6	3	1	0
asparagine	0.222	1 OAA	1	0	2	2	0	0
aspartate	0.647	1 OAA	1	0	0	1	0	0
cysteine	0.015	1 3PG	4	-1	3	1	0	1
glutamate	0.658	1 αKG	1	0	0	1	0	0
glutamine	0.229	1 αKG	1	0	1	2	0	0
glycine	0.632	1 3PG	1	-1	0	1	0	0
glycine[d]	0.632	1 AcCoA	1	-1	0	1	0	0
histidine	0.144	2 R5P,1 3PG	4	-5	16	7	1	0
isoleucine	0.421	1 OAA, 1 PY	5	0	2	1	-1	0
leucine	0.645	2 PY, 1 AcCoA	2	-1	0	1	-2	0
lysine	0.623	1 αKG, 1 AcCoA	3	-1	2	2	-1	0
methionine	0.111	1 OAA, 1 AcCoA	5	0	4	1	0	1
phenylalanine	0.292	1 E4P, 2 PEP	2	0	1	1	-1	0
proline	0.360	1 αKG	2	1	1	1	0	0
serine	0.403	1 3PG	1	-1	0	1	0	0
serine[d]	0.403	1 AcCoA	1	-1	0	1	0	0
threonine	0.416	1 OAA	3	0	2	1	0	0
tryptophane	0.061	1 E4P, 1 PEP, 1 R5P	3	-2	3	2	-1	0
tyrosine	0.222	1 E4P, 2PEP	2	-1	1	1	-1	0
valine	0.577	2 PY	2	0	0	1	-1	0
Nucleotides			NADPH	NADH	~P	NH$_3$	CO$_2$	PO$_4$
ATP	0.754	1 R5P, 1 3PG	3	-3	13	5	1	1
GTP	0.754	1 R5P, 1 3PG	2	-3	15	5	1	1
UTP	1.000	1 R5P, 1 OAA	1	0	7	2	0	1
CTP	0.738	1 R5P, 1 OAA	1	0	8	3	0	1
Lipids			NADPH	NADH	~P			
Lauric	0.674	5 AcCoA	10	0	5			
Palmitoleic	1.000	7 AcCoA	14	1	7			
Oleic	0.403	8 AcCoA	16	1	8			
Glycerol P	0.667	1 TP	0	1	0			
Polysaccharides			NADPH	NADH	~P			
Glycogen	0.741	G6P	0	0	2			
Glucan	0.857	G6P	0	0	2			
Mannan	1.000	G6P	0	0	2			

[a] based on the composition reported by Bruinenberg *et al.* (1983)
[b] key intermediary metabolites.
[c] phosphorylation energy required.
[d] in acetate, glycine and serine are synthesized through a different pathway.
(Fluxes of carbon, phosphorylation, and redox intermediates during growth of *Saccharomyces cerevisiae* on different carbon sources, Cortassa, Aon and Aon, *Biotech. Bioengin.* 47, 193–208. ©copyright 1995 John Wiley & Sons, Inc. Reprinted by permission of Wiley-Liss, Inc., a subsidiary of John Wiley & Sons, Inc.).

Catabolic fluxes

Catabolic reactions account for the provision of energy and reduced intermediates fueling anabolism. A large diversity of catabolic pathways implying different stoichiometries, are known to occur in nature. Among chemoorganotrophs (Fig. 2.1A), a large range of substrates may act as energy sources from the very reduced, e.g. as methane, to oxidized ones, e.g. oxalic acid (Fig. 2.3; see also Table 2.2) (Linton and Stephenson, 1978). The search for microorganisms that degrade xenobiotics has shed light on pathways that use aromatic compounds, e.g. toluene, xylene, or other aromatics, as carbon and energy sources (Wilson and Bouwer, 1997).

Depending on the presence of exogenous or endogenous electron acceptors, aerobic or anaerobic metabolism occurs. Table 2.8 shows the most common catabolic reactions either with oxygen or endogenous C compounds as electron acceptors, and their mass and energy balances. In aerobic metabolism depending on the nicotinamide adenine nucleotide (NADH or FADH) that acts as electron donor, the resulting amount of ATP synthesized will differ (Table 2.8). The resulting P:O ratio, i.e. the mole numbers of ATP synthesized in oxidative phosphorylation per mole atom of oxygen consumed, will be two for FADH, or three for NADH. Thus, the yield of ATP will vary depending on the substrate being oxidized and on the type of redox intermediate synthezised from its oxidation. As an example when succinate is acting as energy source, electrons are transferred via FADH so the maximun P:O ratio is two.

In the 1990s alternative respiration mechanisms have been described according to which a NADH oxidase transfers electrons to molecular oxygen without translocating protons across the mitochondrial inner membrane; thereby, no ATP is synthesized. Thus during microbial growth, the resulting ATP yield, as shown by the P:O ratio, will result from the various stoichiometries of the electron transport pathways.

In the next section we further develop the main concepts about the quantification of energy balance.

Table 2.7 Amounts of precursors required for the synthesis of macromolecular monomers

	Monomer precursor of macromolecules		
	Arginine	Uracil 5' PPP	Glucose (glycogen)
Amount in biomass (mmol g^{-1} dw)	0.187	0.123	0.548
Key intermediary precursor	α-ketoglutarate	Ribose 5P Oxalacetate	Glucose 6 P
Amount of key intermediary precursor	0.187	0.123 (Ribose 5P) 0.123 (Oxalacetate)	0.548
Amount of glucose as carbon source required to obtain the key intermediate	0.187	0.185 $(0.123.1.5)^b$	0.548
NADPH required	0.748	0.123	-
NADH produced	0.187	-	-
Energy as ATP	1.122	0.861	0.548
CO_2 released	-0.187^a	-	-

[a]The CO_2 is in fact required to synthesize oxalacetate from pyruvate
[b]One or 0.5 moles of glucose are required for the synthesis of one mole R5P or oxalacetate, respectively.

Energy Balance

In a bioprocess, the energy balance may be regarded from two different viewpoints: from that of the chemical engineer, who will consider the process as a whole, or from that of the metabolic engineer. The chemical engineer pays special attention to the amount of energy released by the microorganisms. In fact, microorganisms evolve heat during growth and this is relevant because of the importance of temperature control in bioprocesses. Energy is therefore usually provided to cool the bioreactor and various heat exchanger models have been designed. Biochemical engineering textbooks provide further detail.

The metabolic engineer's viewpoint emphasizes the microorganism and its metabolic and physiological behavior. This is the perspective adopted in the present book. Metabolism involves a set of chemical reactions each one with an associated energetic change. These cannot be treated separately from mass balance, even in the black box approach. Since catabolism implies the oxidation of a reduced organic compound, energy is released and ATP synthesis coupled to it. In the first part of this section, we will deal with some general thermodynamic concepts applicable to bioprocesses in general, either via a modular approach to metabolism, or to each of the individual reactions.

Table 2.8. Main catabolic routes rendering ATP and reduction power operative in chemoorganotrophic organisms.

$C_6H_{12}O_6 + 2\ ADP + 2\ NAD^+ + 2\ Pi$	$\Rightarrow$	$2\ Py + 2\ ATP + 2\ NADH$ ([a])
$C_6H_{12}O_6 + ATP + 12\ NADP^+$		$6\ CO_2 + ADP + Pi + 12\ NADPH$ ([b])
$Py + 1\ ADP + Pi + FAD^+ + 4\ NAD^+$	$\Rightarrow$	$3\ CO_2 + ATP + 4\ NADH + FADH$ ([c])
$Py + NADH$	$\Rightarrow$	$EtOH + CO_2 + NAD^+$ ([d])
$Py + NADH$	$\Rightarrow$	$Lactate + NAD^+$ ([e])
$NADH + 3\ ADP + 3\ Pi + 1/2\ O_2$	$\Rightarrow$	$NAD^+ + 3\ ATP + H_2O$ ([f])
$FADH + 2\ ADP + 2\ Pi + 1/2\ O_2$	$\Rightarrow$	$FAD^+ + 2\ ATP + H_2O$ ([g])

[a]Glycolysis via the Embden Meyerhof pathway. The ATP stoichiometry is 1 mole ATP mole^{-1} glucose if the Entner-Doudoroff pathway is used as in some prokaryotes.
[b] Pentose phosphate pathway operating in a cyclic mode to generate reduction equivalents in the form of NADPH to fuel biosynthetic redox reactions.
[c] Tricarboxylic acid overall stoichiometry including the pyruvate dehydrogenase catalyzed step.
[d] Ethanolic fermentation.
[e] Lactic fermentation
[f] Oxidative phosphorylation indicating the mechanistic stoichiometry of 3 from NADH (P:O ratio = 3) The actual stoichiometry is usually lower than the mechanistic one.
[g] Oxidative phosphorylation indicating the mechanistic stoichiometry of 2 from FADH (P:O ratio = 2).

Forms of Energy and Enthalpy

Cellular systems are dissipative in the sense that they sustain fluxes of matter and energy, interconverting the nature of the forces (e.g. chemical, radiant, electrostatic) driving such fluxes. In cells, free energy stores are exchanged essentially between the chemical potentials, electrochemical gradients, and conformational energy of macromolecules.

There are several forms of energy, most of them being interconvertible. These forms of energy are: kinetic, potential, internal, mixing work, flow work and finally heat.

Kinetic energy is associated with movement due to the traslational velocities of the different components of a system. Potential energy is linked to the position of a system in a force field (e.g. gravitational, electromagnetic). In bioprocess engineering, mixing (usually refered to as shaft work) and flow (energy required to pump matter into and out of the system) work, are important terms. Internal energy is associated with molecular and atomic interactions and is measured as changes in the enthalpy of the system. Thermodynamics deals with these energy forms, and a series of state functions that represent them.

The most commonly used state functions are enthalpy and free energy, the former being the most useful to evaluate heat exchange in biotechnological processes. Enthalpy is defined as the sum of the internal energy of a system, U, plus the compression or expansion work, pV:

$$H = U + pV \qquad (2.21)$$

Thermodynamics textbooks provide definitions and examples of enthalpy in detail. The book by Roels "Energetics and Kinetics in Biotechnology" (1983) should be considered as a most useful work for the thermodynamic approach to biotechnological problems.

The general equation of energy balance is equivalent to the one of material balance. It is based on the first law of thermodynamics, and states that in any physico-chemical process (except those involving subatomic reactions), the total energy of the system is conserved:

$$E_{input} - E_{output} = E_{accumul} \qquad (2.22)$$

where E_{input} stands for the flow of energy into the system; E_{output}, the output flow of energy from the system; $E_{accumul}$, the energy accumulated in the system during the process.

More specifically, Eq. 2.22 can be written as follows:

$$\underset{input}{\sum M} (u + e_k + e_p + pv) - \underset{output}{\sum M} (u + e_k + e_p + pv) - Q + W_S = \Delta E \quad (2.23)$$

In most bioprocesses, the contribution of kinetic and potential energy may be considered negligible. Then, Eq. 2.23 reads:

$$\underset{input}{\sum M} h - \underset{output}{\sum M} h - Q + W_S = \Delta E \qquad (2.24)$$

Some special cases of interest are the steady state, or when the system is thermically isolated in an adiabatic device (Eq. 2.25). For instance at the steady state, $\Delta E = 0$ or under adiabatic conditions, $Q = 0$, then Eq. 2.23 becomes:

$$\underset{input}{\sum M} h - \underset{output}{\sum M} h + W_S = \Delta E \qquad (2.25)$$

These thermodynamic functions are called *state properties*. This means that they are only dependent on the state of the system, and that their values are independent of the pathway followed by the system to attain the corresponding state. The latter permits us to compute the value of a state function (e.g. enthalpy) by following a path that allows the use of tabulated state function changes. To exemplify the latter, let us suppose that we are interested in calculating the enthalpy change associated with a particular reaction occurring at high temperature, T_i. A strategy will be to find the tabulated change in enthaply at standard pressure and temperature, and add to it the enthalpy change to cool the reactants from T_i to the standard conditions plus the enthalpy change associated with heating the reaction products to T_i. From a thermodynamic point of view, this procedure is equivalent to the reaction occurring at T_i because enthalpy is a state property.

State functions do not have absolute values, but are relative to a reference state chosen by the experimenter. Conventionally, in biological systems, the reference state corresponds to 25°C temperature, 1 atm of pressure, pure water (i.e. 55.5 M instead of 1 M), and pH 7.0.

Enthalpy changes may occur associated with: *(i)* temperature; *(ii)* physical aspects such as phase changes, mixing or dissolution; *(iii)* reaction.

To compute the change of enthalpy associated with a change in the temperature of the system, a property called "sensible heat" is defined. *Sensible heat* is the enthalpy change associated with a temperature shift. The heat capacity of the system should be known to determine the sensible heat. The specific heat, Cp, which is the heat capacity expressed per unit mass, can be used to calculate the sensible heat as follows:

$$\Delta H = M\, C_P\, (\Delta T) = M\, C_P\, (T_2 - T_1) \tag{2.26}$$

The concepts of *sensible heat* and *heat capacity* are of fundamental relevance for the work with calorimetry, and in their application to unravel microbial energetics (see below).

Enthalpy changes associated with metabolic reactions, can be expressed as the heat of reaction ΔH_{rxn}, as follows:

$$\Delta H_{rxn} = \sum_{products} M\, h - \sum_{reactants} M\, h \text{ (expressed in terms of mass)} \tag{2.27}$$

$$\Delta H_{rxn} = \sum_{products} n\, h - \sum_{reactants} n\, h \text{ (expressed in mole terms)} \tag{2.28}$$

Calorimetric Studies of Energy Metabolism

Calorimetry was initially developed to quantitate heat production during chemical reactions and by microbial fermentations. Actually, the report by Dubrunfaut (1856; quoted by Belaich, 1980) examining bacterial thermogenesis, was the first one on cellular catabolism.

Calorimeters evaluate the amount of heat evolved (or used) by a culture; this equals the ΔH associated with a particular process under observation. The amount of heat evolved, as expected, depends on the nature of the energy source and is proportional to its degree of reduction (Table 2.2). The output of a calorimeter is the rate of heat production (dQ/dt). The latter was found to be proportional to the acceleration of growth rather than to the growth rate itself (Lamprecht, 1980). Thus, thermograms indicate the rate at which metabolism is operating, and the type of substrate breakdown, e.g. anaerobic catabolism evolves much less heat than aerobic catabolism for the same substrate. According to Table 2.9, *S. cerevisiae* releases 139 kJ mol^{-1} substrate growing on glucose on complex media under anaerobic conditions, whereas under aerobiosis the ΔH increases to 2663 kJ mole^{-1} substrate. Large differences were also observed in synthetic media with yeast cells growing on either C-, or N- or C- and N-limited chemostat cultures (Larsson *et al.*, 1993). In the respiro-fermentative mode of glucose breakdown the heat yield, i.e. the heat production per gram of biomass formed, decreased to 7.3 kJ g^{-1} under glucose limitation. The largest value of heat yield, 23.8 kJ g^{-1}, was verified in N-limited cultures at low dilution rates (Larsson *et al.*, 1993).

The respiratory quotient (see Section: *Growth stoichiometry and product formation*) has served as a useful basis for estimation of enthalpy changes in experiments using "indirect calorimetry" with a Warburg apparatus. However, this procedure is subject to experimental error although a first estimation of RQs can be performed. Under fully aerobic conditions, the RQ values for carbohydrates (1.00), fats (0.707), and proteins (0.801), were measured. Since the heat of combustion of these compounds were known (17.2 kJ g^{-1} for carbohydrates, 38.9 kJ g^{-1} for fats and 22.6 kJ g^{-1} for proteins), the enthalpy changes could be readily calculated (reviewed by Lamprecht, 1980). At present, with more sophisticated equipment for both calorimetric and RQ measurements, indirect calorimetry was found to give only a qualitative assessment of enthalpy changes associated with metabolism (Lamprecht, 1980).

As heat production is the result of the whole metabolic machinery, nowadays calorimetry is used in combination with other methods and these enable a thorough interpretation of the energetics of growth and product formation in the analysis of metabolic networks (Larsson *et al.*, 1993).

Heat of Combustion

A useful thermodynamic state function is the heat of combustion. This is defined as the amount of heat released during the reaction of a substance with oxygen to form oxidized products such as CO_2, H_2O and N_2, at standard conditions of temperature and pressure, i.e. $25°C$, 1 atm.

The standard heat of combustion is used to estimate the reaction enthalpy changes, as follows:

$$\Delta H_c^o = \sum_{reactants} n\, \Delta h_c^o - \sum_{products} n\, \Delta h_c^o \qquad (2.29)$$

Indeed, the growth of a microorganism is related to its physiology and bioenergetics. The growth yield of microorganisms on a given substrate, and the heat of combustion of that substrate, were found to be linearly related (Fig. 2.3). The heat of combustion was interpreted as a measure of the carbon and energy content of an organic substrate.

Figure 2.3 shows that in the range of 2.5 to 11.0 kcal g^{-1} substrate-C, the enthalpy of combustion and the growth yield (ranging from 0.2 to 1.4 g dw g^{-1} substrate-C) display a linear correlation. At enthalpies of combustion higher than 11.0 kcal g^{-1} substrate-C, growth is no longer energy-limited, and an upper limit in the relationship between Y_{XS} and ΔH_c^o is attained.

The maximal Y_{XS} of 1.43 g dw g^{-1} substrate-C, was interpreted as the achievement of a stoichiometric limit. This relationship is also discussed in the context of the work of Sauer *et al.* (1998) as applied to the growth and metabolite production by *Bacillus subtilis*, and analyzed with a stoichiometric model (see Chapter 4, Section: *A comparison between different methods of metabolic flux analysis*).

Stouthamer and van Verseveld (1985) have redrawn the data of Linton and Stephenson (1978) on the bases of the degree of reduction of the growth substrates. They introduced the concept that during growth of organisms, adaptation of the energy generation mechanisms occur according to ATP demand by biomass synthetic processes. Such adaptation can be accomplished through changes in the P:O ratio (deletion of one or more phosphorylation sites in the respiratory chain) and occurrence of energy spilling mechanisms (Stouthamer and van Verseveld, 1985). This explains why the growth of *Pseudomonas denitrificans* with mannitol or methanol is energy-limited.

Table 2.9. Calorimetrically determined parameters of growing yeast cultures under varying conditions. Enthalpy changes ΔH (kJ mol^{-1} substrate specific rate of heat production $d\Delta h/dt$ (W g^{-1}), enthalpy change per formed biomass AH' (kJ g^{-1}), growth yield Y (g dry weight g^{-1} substrate). Some values are calculated from the data in the literature. AN, anaerobic; (AN), nearly anaerobic; AE, aerobic (AE), nearly aerobic.

Yeast	Condition	Medium	Substrate	ΔH	$d\Delta h/dt$	ΔH	Y
a. Anaerobic batch cultures							
Brewer's yeast	AN	Complex	Sugar	98.4	-		
Yeast (undefined)	(AN)	Complex	Cane sugar	2 x 110	-	2 x 27.9	0.0218
Yeast (undefined)	(AN)	Complex	Maltose	2 x 112	-	2 x 32.1	0.0193
S. cerevisiae	AN	Synthetic	Glucose	96.3	-	6.28	0.085
Baker's yeast	AN	Complex	Glucose	82.5	-	1.93	0.237
S. cerevisiae Y Fa	AN	Complex	Glucose	119	-	5.25	0.126
S. cerevisiae 211	(AN)	Complex	Glucose	99.2	-	4.23	0.131
S. cerevisiae 211	(AN)	Complex	Glucose	89.6	0.17	4.48	0.111
S. cerevisiae	(AN)	Complex	Glucose	63.9	-	-	-
Baker's yeast	AN	Complex	Glucose	129	0.34	-	-
S. cerevisiae	AN	Complex	Fructose	124	-	4.78	0.144
S. cerevisiae	AN	Complex	Galactose	125	-	4.82	0.144
S. cerevisiae	AN		Maltose	2 x 126	-	2 x 6.31	0.111
b. Aerobic batch cultures							
Rhodotorula sp.	AE	Complex	Glucose	536		1.05	7.61
C. intermedia	(AE)		Glucose	1695		11.4	0.826
S. cerevisiae 211	AE	Solid medium	Glucose	1674		-	-
Baker's yeast	AE		Glucose	-	1.97	-	-
Kl. Aerogenes	AE	Synthetic	Glucose	-	3.0~4.2	-	0.19
D. hansenii	AE	Synthetic	Glucose	1130-169	0.4	12.6-18.8	0.5
S. cerevisiae	AE	Complex	Ethanol	854	-	33.7	0.55
Baker's yeast	AE	Complex	Ethanol	-	1.23	-	-
Hansenula sp.	AE		Ethanol	544	2.09	16.7	0.61
S. cerevisiae	AE	Complex	Acetic acid	678	-	45.2	0.25
c. Continuous cultures							
S. cerevisiae	AN	Complex	Glucose	139	0.2	10.4	0.074
S. cerevisiae	AE	Complex	Glucose	2663	2.15	14.5	1.017
KI. Aerogenes	AE	Synthetic	Glucose	-	0.13	-	-
S. cerevisiae	AE	Synthetic	Glucose	1070	1.54	10.4	0.572
S. cerevisiae	AE	Synthetic	Ethanol	616	-	28.3	0.474
Hansenula sp.	AE		Ethanol	607	-	18	0.73

Table 2.2 shows the proportionality between the enthalpy and free energy of
combustion of organic compounds (expressed on a carbon basis) and their degree
of reduction. Among the organic compounds analyzed, are a number of carbon
and energy substrates able to sustain microbial growth as well as compounds that
take part of the macromolecular composition of cells, e.g. amino acids. The
above mentioned proportionality is evidenced through the constancy in the ratio
between the heat of combustion over the degree of reduction, with an average
proportionality constant of 115 kJ C-mole^{-1} (Table 2.2). The data shown in Table
2.2, allow to write the following formula to compute the molar heat of
combustion of any organic compound:

$$\Delta h_c^o = -q\, \gamma x_C \tag{2.30}$$

where q is the heat evolved per mole of available electrons transferred to oxygen
during combustion; γ, the degree of reduction of the compound with respect to
N_2, and x_C, the number of C atoms in the molecular formula.

Usually, biotechnological processes involve microbial growth. The growth of
a microorganism results from a complex set of chemical reactions as well as
regulatory networks (see Fig. 1.4, Chapter 1). The energy or heat evolved by
microbial growth may be computed through the heat of combustion of the
various chemicals involved, carbon and nitrogen substrates or biomass, as
follows:

$$C_wH_xO_yN_z + a\,O_2 + b\,H_gO_hN_i \Rightarrow c\,CH_\alpha O_\beta N_\delta + d\,CO_2 + e\,H_2O \tag{2.31}$$

Growth can occur either under aerobic or anaerobic conditions depending on
the electron acceptor used. Equation 2.29 corresponds to growth with oxygen as
electron acceptor, while Eq. 2.32 represents growth with endogenous organic
compounds as electron acceptors (anaerobic growth):

$$C_wH_xO_yN_z + b\,H_gO_hN_i \Rightarrow c\,CH_\alpha O_\beta N_\delta + d\,CO_2 + e\,H_2O + f\,C_jH_kO_lN_m \tag{2.32}$$

In all cases, the heat of growth may be calculated from the following
expression:

$$\Delta H_{rxn}^o = n\,(\Delta h_c^o)_{substrate} + n\,(\Delta h_c^o)_{NH3} - n\,(\Delta h_c^o)_{biomass} - n\,(\Delta h_c^o)_{product} \tag{2.33}$$

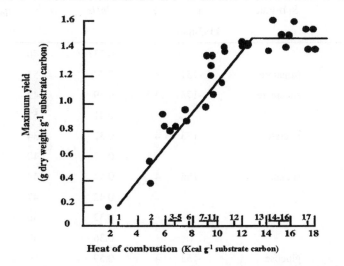

Figure 2.3. Maximum growth yield relationship with enthalpy of combustion of various organic compounds supporting microbial growth. The maximum growth yields observed in a wide range of organic substrates for various microorganisms growing in batch or continuous cultures are plotted against the heat of combustion as a measure of the energy content of the compound and directly related to its degree of reduction (Table 2.2). 1, oxalate; 2, formate; 3, citrate; 4, malate; 5, fumarate; 6, succinate; 7, acetate; 8, benzoate; 9, glucose; 10, phenylacetic acid; 11, mannitol; 12, glycerol; 13, ethanol; 14, propane; 15, methanol; 16, ethane; 17, methane.
(Reprinted from *FEMS Microbiol. Lett.* 3, 95-98, Linton, J.D. and Stephenson, R.J. A preliminary study on growth yields in relation to the carbon and energy content of various organic growth substrates. ©copyright 1978, with permission from Elsevier Science).

Table 2.10. Efficiencies of microbial growth under aerobic conditions on substrates of different degree of reduction. The values of standard Gibbs free energy of the substrate, its degree of reduction, the growth yield of the indicated organism on that substrate (fifth column) and the efficiencies in column (η_{th}^R) were taken from Westerhoff *et al.* (1982). The η_{th}^{Wff} was calculated with the following formula:

$$\eta_{th}^{Wff} = \frac{\eta_{th}^R - (J_b / J_s)}{1 - (J_b / J_s)}$$

Organism	Substrate	μ^*	γ	Jb/Js	η_{th}^R	η_{th}^{Wff}
		kJ C-mol^{-1}				
Pseudomonas oxalaticus	oxalate	-337	1	0.07	24	20
	formate	-335	2	0.18	34	20
Pseudomonas sp.				0.18	34	20
Pseudomonas denitrificans	citrate	-195	3	0.38	54	26
A. aerogenes				0.34	49	23
Ps. denitrificans	malate	-212	3	0.37	52	24

Table 2.10 (*Continued*)

Organism	Substrate	μ* kJ C-mol^{-1}	γ	Jb/Js	η_{th}^R	η_{th}^{Wff}
Ps. fluorescens				0.33	46	19
Ps. denitrificans	fumarate	-151	3	0.37	53	25
Ps. denitrificans	succinate	-173	3.5	0.39	49	16
Pseudomonas sp				0.41	52	19
A. aerogenes	Lactate	-173	4	0.32	34	3
Ps. fluorescens				0.37	40	5
Pseudomonas sp	acetate	-186	4	0.44	49	9
Candida utilis				0.42	47	9
Ps. fluorescens				0.32	36	6
C. tropicalis				0.36	40	6
S. cerevisiae	glucose	-153	4	0.59	59	0
S. cerevisiae				0.57	57	0
E. coli				0.62	62	0
Penicillium chrysogenum				0.54	54	0
A. acogen	glycerol	-163	4.67	0.66	57	-26
C. tropicalis	ethanol	-91	6	0.61	44	-44
C. boidinii				0.61	44	-44
C. utilis				0.61	44	-44
Ps. Fluorescens				0.43	23	-35
C. utilis				0.55	39	-36
C. brassicae				0.64	46	-50
C. boidinii	Methanol	176	6	0.52	36	-33
Klebsiella sp				0.47	32	-28
M. methanolica				0.6	41	-48
Candida N-17				0.46	31	-28
H. polymorpha				0.45	31	-25
Pseudomonas C				0.67	46	-64
				0.5	34	-31
M. methanolica				0.64	44	-56
C. tropicalis	hexadecane	5	6.13	0.56	41	-34
C. lipolytica	dodecane	4	6.17	0.41	30	-19
Job 5	propane	-8	6.67	0.71	48	-79
	ethane	-16	7	0.71	46	-82
M. capsulatus	methane	-51	8	0.63	37	-70
M. methanooxidans				0.68	40	-88

In microbiological processes, it is usual that only the heat of reactions, phase change, and shaft work, are employed to calculate the energy balance, whereas other energy forms are frequently negligible.

So far, only the general thermodynamic concepts have been discussed. In the next section we open the "black box".

An Energetic View of Microbial Metabolism

The main aim of this section is to visualize the sinks of energy during microbial growth. The energy demands during microbial growth are associated with:

- Synthesis of precursors of cellular material.

- Polymerization and acquisition of secondary and tertiary structure of proteins (folding), DNA and RNA.

- Formation of supramolecular assemblies (enzyme-enzyme complexes, membranes, ribosomes)

- Assembly of cell organelles and the whole cell.

Other energy-demanding processes taking place during cell growth, but not directly associated with biomass synthesis, are:

- Maintenance of cell physiological status (intracellular pH, ionic balance, osmotic regulation).

- Transport of substances to and from the cell.

The energetic yield of a microbial process may be directly calculated from the known biochemistry of catabolic pathways (e.g. glycolysis occurring through the Embden-Meyerhoff or Entner-Doudoroff pathways, and the efficiency of oxidative phosphorylation through the P:O ratio).

A useful quantity regarding the energetics of microbial growth is the biomass yield based on ATP, Y_{ATP}, that has been extensively analyzed by Stouthamer (1979). This can be exactly calculated only during anaerobic growth, because the efficiency and branching of the aerobic respiratory chain are difficult to assess experimentally (Stouthamer and van Verseveld, 1985). Y_{ATP} values are influenced by a number of factors, e.g. the presence of exogenous electron acceptor, the carbon source and the complexity of the medium, the nature of the nitrogen source and the chemical composition of the organism under specific conditions. Regarding the latter it has been found that organisms exhibiting large differences in macromolecular compositions showed little variance in Y_{ATP}^{max}, e.g. 28.8 against 25 (g cells mol^{-1} ATP) for *E. coli* and *Aerobacter aerogenes*, respectively. On the other hand the nature of the carbon or the nitrogen sources and especially the metabolic pathway through which they are assimilated exerted a strong influence on the ATP yield (Stouthamer, 1979).

It has been calculated that only between 15 to 25% of the total energy demand, was required for the synthesis of macromolecular precursors (Lamprecht, 1980). A large proportion of total energy expenditure is required to maintain cellular organization. This implies that multicellular organisms with a higher degree of complexity require even a larger energy supply to maintain that organization. Calorimetric determinations have revealed that yeast cells release 157-fold higher amounts of heat than a human; the comparison made on a nitrogen content basis. In turn, a horse releases three times as much heat as a human (quoted in Lamprecht, 1980).

Bioenergetics regards cellular metabolism as an energy-transducing network that couples the release of energy from the oxidation of organic compounds to the synthesis of cellular biomass and other energy-demanding processes. Within this framework, a series of thermodynamic functions have been defined to understand cellular optimization principles as a function of fitness (Westerhoff and van Dam, 1987). These authors analyzed microbial growth (e.g. bacterial, fungal) on various substrates. Thermodynamic efficiency was calculated according to two different definitions (Table 2.10) (Roels, 1983; Westerhoff and van Dam, 1987). According to Westerhoff and van Dam (1987), thermodynamic efficiency becomes negative when the energy content of the substrate is so high that additional energy provided by catabolism (ATP and reducing equivalents) contribute to fueling the microbial system just to make it run faster (Westerhoff *et al.*, 1982). Reasoning in terms of Steps I and II of anabolism (see Section: *Anabolic fluxes*), a substrate with a high energy content generates in the pathways involved in Step I a surplus of ATP sufficient to fuel Step II, together with all other energy-demanding processes such as polymerization or assembly of cellular structures. Efficiencies lower than zero are obtained with substrates more reduced than biomass. When microbial growth occurs on C substrates with a similar reduction degree as biomass, e.g. glucose, lactate or acetate, the thermodynamic efficiency of growth is almost zero. Substrates that are more oxidized than biomass, e.g. formate, oxalate, sustain maximal efficiency values of 25% (Table 2.10). In fact, the theoretical maximum of thermodynamic efficiency (i.e. 100 %) is never attained, since this would imply that there are no flows through the system, i.e. the growth rate would be null in such a case (Westerhoff and van Dam, 1987).

Chapter 3

Cell Growth and Metabolite Production. Basic Concepts

Microbial Growth under Steady and Balanced Conditions

Microbial populations grow in geometric progression. The dynamic states of cultures of microorganisms are essentially of two sorts: transient or steady, irrespective of their state of growth or quiescence. The main difference between transient or steady state is their time-dependence (see Chapter 5). In the steady state, two kinds of behavior may be described: non-expanding (steady state) and expanding (balanced growth) ones. The former are obtained in chemostat cultures whereas the latter occur in batch logarithmic growth.

Growth is balanced when the specific rate of change of all metabolic variables (concentration or total mass) is constant (Barford *et al.*, 1982):

$$\frac{1}{x_{iAV}} \frac{\Delta x_i}{\Delta t} = constant \tag{3.1}$$

where Δx_i is the change in x_i during time Δt; x_{iAV} is the average value of x_i during time Δt.

A steady state is a sort of balanced growth, but with the further requirement that the overall rate of change of any metabolic variable or biomass be zero:

$$\frac{dx_i}{dt} = 0 \tag{3.2}$$

Analysis of a growing population, at any time, reveals that on average, all cells process the available nutrients, produce new biomass and increase in cell size at exactly the same rate. By definition, the time taken to go from one generation (n-1) to the next generation (n) is constant under stable environmental conditions, and so the rate of increase in cell number in the population increases with time. In fact, the population number increases exponentially, and the rate of acceleration in the population size is governed by two major factors. Firstly, the rate is controlled by the intrinsic nature of the organism that is its basic genetic structure and potential, which largely determine its physiological characteristics. Secondly, the rate of population increase is substantially modified by extrinsic factors due to the organism's growth environment (Slater, 1985). So in 'rich' environments containing many preformed cellular components, such as amino acids and nucleic acids, the rate of increase is more rapid than in a 'poor' environment in which the growing organism has to synthesize all its cellular requirements, devoting proportionally much more of the available resources and energy to these activities. In both rich and poor environments, the rate of increase of cell numbers per generation time are constant, but the time taken to complete each generation varies. It is much shorter in a rich environment compared with that in a poor environment (Slater, 1985).

Thus the rate of cell number increase with time is a variable dictated by the type of organism and the nature of its growth environment. Under constant environmental conditions the time taken to complete each generation and double the size of the population is a characteristic constant known as the culture doubling time t_d.

The rate of change of the population size, or more conveniently the rate of growth, is directly proportional to the initial population size,

$$\frac{dx_i}{dt} = \mu\, x_o \qquad (3.3)$$

In Eq. 3.3 μ is a proportionality constant known as the specific growth rate with a precise biological meaning since it is a measure of the number of new individuals produced by a given number of existing individuals in a fixed period of growth time. If the rate of growth is maximized, then so to is μ, which becomes the maximum specific growth rate μ_{max}. In many closed (batch) cultures the ideal conditions are achieved and growth proceeds at μ_{max}.

By integration from $t = 0$ to $t = t$ Eq. 3.3 has the solution:

$$x_i = x_o\, e^{\mu t} \qquad (3.4)$$

which describes an exponential curve, which in linear form reads:

$$ln\ x_t\ =\ ln\ x_o\ +\ \mu t \qquad (3.5)$$

its slope is:

$$\mu = \frac{0.693}{t_d} \qquad (3.6)$$

Thus, the specific growth rate is inversely proportional to the culture doubling time.

The basic growth Eq. 3.4 may be derived either by considering populations as numbers of individual cells, or rates of increase as increases in the number of cells in the population (see Slater, 1985).

Unlimited quantities of all growth resources might exist only for short periods of time (e.g. early stages of a batch culture), but in most natural habitats, resource limitation is the rule rather than the exception. Thus growth limitation by substrate depletion (of the growth limiting substrate) restricts the final size of the population, and indeed the two parameters may be shown to be proportional to each other:

$$xf \propto S_r \qquad (3.7)$$

where S_r is the initial concentration of the growth-limiting substrate, and xf is the final population size. Thus:

$$xf = Y \, S_r \qquad (3.8)$$

where, Y is a proportionality constant known as the observed growth yield. In its simplest form it is defined as that weight of new biomass produced as a result of utilizing unit amount of the growth-limiting substrate. The yield term indicates how much of the available substrate is used for new biomass production and also the amount used for energy generation to drive biosynthetic reactions (see Chapter 2). Table 3.1 shows the biomass growth yield on the bases of glucose or ATP for several organisms growing anaerobically (Stouthamer, 1979). That a relationship exists between the amount of growth of a microorganism and the amount of ATP that could be obtained from the energy source in the medium, was first suggested by Bauchop and Elsden (1960). However, this relationship has turned out to be more complex than the originally suggested direct proportionality (Stouthamer, 1979; Stouthamer and van Verseveld, 1985). Thus, both theoretical and experimental data show that Y_{ATP} is not a biological constant (Table 3.1), and that the ATP requirement for biomass formation depends on the carbon substrate, the efficiency of oxidative phosphorylation, the anabolic and catabolic pathways utilized (see Table 4.1), the presence of preformed monomers, and the nitrogen source (Stouthamer, 1979; Stouthamer and van Verseveld, 1985). Moreover, measured values are mostly 50-60% of the theoretical calculated ones. Several possible explanations have been offered to understand this phenomenon (Tempest and Neijssel, 1984; Stouthamer and van Verseveld, 1985; Verdoni *et al.*, 1990).

Assimilatory and dissimilatory flows are needed to incorporate carbon substrates into biomass. The assimilatory flow is defined as the substrate flux directed towards biomass synthesis (anabolism), whereas the dissimilatory flow is the amount of substrate needed to release energy for other purposes such as transport, futile cycles and synthesis of biomass (Gommers *et al.*, 1988). Clearly, the greater the amount of energy generation which is diverted for biosynthetic purposes, the less the amount of available substrate which has to be dissimilated, and the higher the yield (see Table 4.1).

During the growth and decline phases of batch culture, the specific growth rate of cells is dependent on the concentration of nutrients in the medium. Often, a single substrate exerts a dominant influence on rate of growth; this component is known as the growth-limiting substrate. The growth-limiting substrate is often the carbon or nitrogen source, although in some cases it is oxygen or another oxidant such as nitrate. Monod was the first to establish the relationship between growth-limiting substrate concentration and specific growth rate, by recognizing that it follows a rectangular hyperbola and closely resembles that between the velocity of an enzyme-catalyzed reaction and substrate concentration. Monod deduced the following expression for microbial cultures:

$$\mu = \frac{\mu_{max}\,[S]}{K_S + [S]} \tag{3.9}$$

In Eq. 3.9 *[S]* is the concentration of growth-limiting substrate, and K_S is the saturation constant. The saturation constant is a measure of the affinity the organism has for the growth-limiting substrate. Typical values of K_S are shown in Table 3.2; they are very small, of the order of mg per litre for carbohydrate and µg per litre for other compounds, such as amino acids (Doran, 1995).

Table 3.1. Y_{ATP} for batch cultures of a number of microorganisms growing anaerobically in various media on glucose.

Organism	$Y_{X/C}$ (g dw mol^{-1} substrate)	Y_{ATP} (gmol gmol^{-1} substrate)
Streptocoecus faecalis	20.0-37.5	10.9
Streptococcus agalactiae	20.8	9.3
Streptococcus pyogenes	25.5	9.8
Lactobacillus plantarum	20.4	10.2
Lactobacillus casei	42.9	20.9
Bifidobacterium bifidum	37.4	13.1
Saccharomyces cerevisiae	18.8-22.3	10.2
Saccharomyces rosei	22.0-24.6	11.6
Zymomonas mobilis	8.5	8.5
	6.5	6.5
	4.7	4.7
Zymomonas anaerobia	5.9	5.9
Sarciiza ventriculi	30.5	11.7
Aerobacter aerogenes	26.1	10.2
	47	18
	69.5	28.5
Aerobacter cloacae	17.7-27.1	11.9
Escherichia coli	25.8	11.2
Ruminococcus flavefaciens	29.1	10.6
Proteus mirabilis	14	5.5
	38.3	12.6
	48.5	18.6
Actinomyces israeli	24.7	12.3
Clostridium perfringens	45	14.6
Streptococcus diacetilactis	35.2	15.6
	43.8	21.5
Streptocoecus cremoris	31.4	13.9
	38.5	18.9

Reproduced from Stouthamer, 1979, *Int. Rev. Biochem.* 21, 1-47.

The level of the growth-limiting substrate in culture media is normally much greater than K_S (usually $[S] > 10\ K_S$). This explains why μ remains constant and equal to μ_{max} in batch culture until the medium is virtually exhausted of substrate (see also below and Chapter 5 for the treatment of this subject in open, chemostat, systems).

From Eqs. 3.7 and 3.8 it holds that in a small time interval, δt, the population increases in size by a small amount, δx, as a consequence of using a small amount of the growth-limiting substrate -δs. Thus:

$$-\frac{dx}{ds} = Y \tag{3.10}$$

that is the rate of change of biomass with respect to used substrate depends on the observed growth yield. Hence, the rate of change of the growth-limiting substrate may be calculated by substituting Eq. 3.3 for dx and rearranging:

$$-\frac{d[S]}{dt} = \frac{\mu\, x}{Y} \tag{3.11}$$

the negative sign indicates a decrease in substrate concentration with time. The constants μ and Y may be replaced by another constant q, the specific metabolic rate:

$$-\frac{d[S]}{dt} = q\, x \tag{3.12}$$

The specific metabolic rate defines the rate of uptake of the growth-limiting substrate by a unit of amount of biomass per unit time, and can vary widely, since μ and Y are subject to environmental variation.

In heterogeneous systems, where the rates of reaction and substrate mass transfer are not independent, the rate of mass transfer depends on the concentration gradient established in the system which in turn depends on the rate of substrate depletion by reaction. Under those conditions, the rate of substrate uptake and the affinity constant of the substrate K_S by the cell become very important.

Considering that the rate of substrate S depletion by a microorganism immobilized in a spherical particle is:

$$r_s = r_s^{max} V_p \frac{[S]}{K_S + [S]} \qquad (3.13)$$

where r_s are moles of S consumed per unit time; Ks, affinity constant; Vp, particle volume; r_s^{max}, maximal rate of S consumption inside the particle expressed in terms of S concentration. When the rate of transport of S to the particle is greater than r_s, the process is kinetically-controlled. Diffusional control occurs when the rate of transport of S to the particle is lower than r_s. The metabolism and physiology of the microorganism become limiting when the process is kinetically-controlled. In several cases, compelling evidence shows that the rate of substrate uptake by the cell is a main rate-controling step of yeast catabolism (Aon and Cortassa, 1994, 1997, 1998).

Table 3.2. Effects of substrate concentration on growth rate. Ks values for several organisms (Data were taken from Pirt, 1975, and Wang *et al.*, 1979).

Microorganism (genus)	Limiting Substrate	Ks (mg 1^{-1})
Saccharomyces	Glucose	25
Escherichia	Glucose	4.0
	Lactose	20
	Phosphate	1.6
Aspergillus	Glucose	5.0
Candida	Glycerol	4.5
	Oxygen	0.042-0.45
Pseudomonas	Methanol	0.7
	Methane	0.4
Klebsiella	Carbon dioxide	0.4
	Magnesium	0.56
	Potassium	0.39
	Sulfate	2.7
Hansenula	Methanol	120.0
	Ribose	3.0
Cryptococcus	Thiamine	1.4×10^{-7}

Microbial Energetics under Steady State Conditions

Monod defined the macroscopic yield of biomass, or observed growth yield (see Eq. 3.8) as the ratio of the biomass produced to substrate consumed. He described the dependence of the growth rate on the concentration of the growth-limiting substrate (see Eq. 3.9). Following the introduction of continuous cultivation techniques, it was shown that in carbon-limited continuous cultures the growth yield, Y_S, was not constant but decreased as the dilution rate, D (=growth rate), decreased (see Slater, 1985, and refs. therein). This effect was attributed to what he called the *endogenous metabolism*. Thus during growth the consumption of the energy source is partly growth-dependent and partly growth-independent. Assuming the latter, the following equation can be derived that relates the growth yield and specific growth rate (Pirt, 1975; Stouthamer, 1979):

$$\frac{1}{Y_{xs}} = \frac{1}{Y_{xs}^{max}} + \frac{m_S}{\mu} \tag{3.14}$$

where Y_{xs} is the yield of biomass, x, on substrate, s; m_s, the maintenance coefficient (mol of substrate per g dry weight per hr), and Y_{xs}^{max} is the yield biomass after correction for energy of maintenance (in g dry weight per mole of substrate). In chemostat cultures, the maximum growth yields for substrate (Y_{xs}^{max}) and oxygen (Y_{o2}^{max}) may be obtained from the plots of q_s and q_{o2} versus D by the following equation (Stouthamer, 1979; Verdoni *et al.*, 1992):

$$q_S = \frac{1}{Y_{xs}^{max}} D + m_S \tag{3.15}$$

A similar reasoning may be followed for the determination of the maximum growth yield based on oxygen (Y_{o2}^{max}) or ATP (Y_{ATP}^{max}) from a plot of q_{o2} or q_{ATP} versus D (Stouthamer, 1979; Verdoni*et al.*, 1992).

As an example, the specific rate of ATP production, q_{ATP} (in mmol ATP h^{-1} g [dry weight]$^{-1}$) in oxygen-limited chemostat cultures of *P. mendocina* growing on glucose, was calculated as follows (Verdoni *et al.*, 1992):

$$q_{ATP} = q_{glc} (1-\beta) + q_{AA} + q_{o2} 3 (P/O) + 3 q_{alg} \tag{3.16}$$

where q_{AA} and q_{alg} are the specific rates of acetate and alginate production, respectively; the number of ATP molecules formed by substrate phosphorylation during the complete oxidation of the substrate is taken into account by q_{glc} (one

ATP) and q_{AA} (one ATP). The fermentation of glucose to acetic acid by the Entner-Doudoroff pathway that operates in *P. mendocina* yields 2 mol of ATP per mol of glucose (Lessie and Phibbs, 1984); β is the part of the substrate that is assimilated. The value of β may be calculated as the ratio between Y_{glc}^{max} values measured under oxygen-limited conditions ($Y_{glc}^{max}=70$) and the molecular weight of *P. mendocina* according to its chemical composition (molecular weight=152.4; see Chapter 2) (Verdoni *et al.*, 1992).

The coefficient 3 that affects the q_{alg} term represents the net yield in ATP as a result of alginate synthesis. It is estimated that for alginate, 3 ATP equivalents are utilized for each uronic acid monomer incorporated (see Fig. 7.2 in Chapter 7) (Jarman and Pace, 1984; Verdoni *et al.*, 1992). At P/O ratios of 3, the alginate biosynthetic pathway becomes net ATP yielding, the 3 ATP equivalents required per monomer polymerized being supplied by the oxidation of the 2 NAD(P)H (6 ATP produced when oxidized via oxidative phosphorylation) generated from uronic acid synthesis (Jarman and Pace, 1984; Verdoni *et al.*, 1992).

Growth Kinetics under Steady State Conditions

Open growth systems differ from closed ones in that there is a continuous input of growth substrates and removal of waste products, cells and unused substrates (see Chapter 2). These systems known as continuous-flow cultures, enable the exponential growth phase to be prolonged indefinitely, establishing steady state conditions. Continuous-flow cultures present additional advantages: specific growth rate may be directly set by the experimenter, substrate-limited growth may be established, submaximal growth rates can be imposed, and biomass concentration may be set independently of the growth rate (Pirt, 1975; Slater 1985).

The most widely used continuous culture system is the chemostat, characterized by growth control through a growth-limiting substrate. The use of chemostat or continuous cultures to study a microorganism at the steady state, provides a rigorous experimental approach for the quantitative evaluation of microorganism's physiology and metabolism (see Chapters 1, 4, 5). Besides, chemostat cultures allow the definition of the phase of behavior which suits a purpose we may decide upon, e.g. maximum substrate consumption and output fluxes of metabolic by-products of interest (see Chapter 4). The behavior of the steady state fluxes as a function of a parameter in the chemostat is, conceptually, the same to that described by a bifurcation diagram (Aon and Cortassa, 1997; see Chapter 5).

The Dilution Rate

In a chemostat, the concentration of the growth-limiting substrate clearly depends on the rate at which the organisms use it and on its rate of supply. The growth-limiting substrate concentration depends on a ratio, known as the dilution rate, D, of the flow rate F through the system and the volume of the culture, V, in the vessel: $D = F / V D$, has units of reciprocal time (h^{-1}), and is a measure of the number of volume changes achieved in unit time.

The Dilution Rate and Biomass Concentration

If it is assumed for a moment that the organisms are non-growing, then the rate of change of the organism concentration in a system with D is given by:

$$\frac{dx}{dt} = -D x \tag{3.17}$$

and by integration:

$$x_i = x_o e^{-Dt} \tag{3.18}$$

Equation 3.18 describes exponential decay and constitutes a measure of the culture washout.

In a growing culture, an additional factor influencing the rate of change of biomass concentration is a growth term, such that:

$$\frac{dx}{dt} = \mu x - D x \tag{3.19}$$

$$\frac{dx}{dt} = [\mu - D] x \tag{3.20}$$

$$\frac{dx}{dt} = \left[\frac{\mu_{max} [S]}{K_S + [S]} - D \right] x \tag{3.21}$$

Equation 3.21 predicts three general cases for the overall rate of change of biomass concentration (Pirt, 1975; Slater, 1985) (Fig. 3.1 top panel). *(i)* If $\mu > D$, dx/dt is positive; the biomass concentration increases and the rate of biomass production is greater than the rate of culture washout (Fig. 3.1 top panel, curve 3). *(ii)* If $\mu < D$, dx/dt is negative; the biomass concentration declines and the growth rate is less than the washout rate (Fig. 3.1 top panel, curve 1). *(iii)* If $\mu = D$, dx/dt is zero and the rate of biomass production balances the rate of culture washout. Under these conditions the culture is said to be in a steady state (Fig. 3.1 top panel, curve 2). This is the preferred stable state and in time all chemostat cultures will reach a steady state provided that $D < \mu_{max}$ and the environmental conditions are kept constant.

The Dilution Rate and the Growth-limiting Substrate Concentration

A similar balanced equation may be derived to describe the change of growth-limiting substrate, such that:

$$\frac{d[S]}{dt} = D[S_r] - D[S] - \frac{\mu x}{Y} \tag{3.22}$$

$$\frac{d[S]}{dt} = D([S_r] - [S]) - \frac{\mu x}{Y} \tag{3.23}$$

The first two terms are rates dependent on D, whilst the growth term was derived previously (Eq. 11]. The two terms $([Sr]—[S])$ and x/Y are equivalent, and both measure the proportion of substrate used for growth, S_g, giving equation:

$$\frac{d[S]}{dt} = D[S_g] - \mu x_g \tag{3.24}$$

Three general cases may be considered (Pirt, 1975; Slater, 1985) (Fig. 3.1 bottom panel). (i) If $D > \mu$, $d[S]/dt$ is positive; the growth-limiting substrate concentration increases and the rate of supply of growth-limiting substrate is less than the rate of use by x (Fig. 3.1 bottom panel, curve 1). (ii) If $D < \mu$, $d[S]/dt$ is negative, and the overall growth-limiting substrate concentration declines (Fig. 3.1 bottom panel, curve 3). (iii) If $D = \mu$, $d[S]/dt$ is zero and the growth-limiting substrate concentration in the culture vessel reaches a steady state (Fig. 3.1 bottom panel, curve 2). This occurs at the same time as the value for $dx/dt = 0$ (Fig. 3.1 top panel).

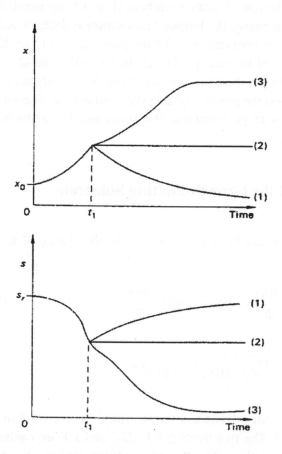

Figure 3.1. The three possible outcomes of a chemostat culture in which growth rate of the biomass (*x*) is limited by the concentration of the growth-limiting nutrient (*s*). The flow of medium containing growth-limiting nutrient at concentration *s*, is started at time, *t₁*. The different cases are: *(i)* rate of wash out of biomass exceeds maximum growth rate; *(ii)* rate of wash out of biomass=maximum growth rate; *(iii)* initial rate of wash out is less than maximum growth rate of biomass (Reproduced from Pirt, S.J. (1975) *Principles of Microbe and Cell Cultivation.* ©copyright Blackwell Scientific Public).

Biomass and Growth-limiting Substrate Concentration at the Steady State

Chemostat culture systems establish steady state conditions when $d[S]/dt = 0$ and $dx/dt = 0$. These values are attained for unique values of x and $[S]$ for a given value of D and constant environmental conditions (Fig. 3.2).

By substituting in Eq. 3.21 and then rearranging, Eq. 3.27 is obtained as follows:

$$0 = \left[\frac{\mu_{max} [S]}{K_S + [S]} - D \right] x \tag{3.25}$$

$$D = \frac{\mu_{max} [S]}{K_S + [S]} \tag{3.26}$$

$$[\bar{S}] = \frac{K_S D}{\mu_{max} - D} \tag{3.27}$$

Similarly, Eq. 3.23 leads to Eq. 3.29.

$$0 = D \left([S_r] - [\bar{S}] \right) - \frac{\mu \bar{x}}{Y} \tag{3.28}$$

$$\bar{x} = Y \left([S_r] - [\bar{S}] \right) \tag{3.29}$$

Substituting Eq. 3.27 into Eq. 3.29 gives Eq. 3.30.

$$\bar{x} = Y \left([S_r] - \frac{K_S D}{\mu_{max} - D} \right) \tag{3.30}$$

Equations 3.27 and 3.30 enable the growth-limiting substrate and steady state biomass concentrations, respectively, to be calculated provided that three basic growth parameters, Ks, μ_{max} and Y, are known. Since for a given organism under constant conditions these parameters are constant, then the unused growth-limiting substrate concentration in the culture vessel depends solely on the imposed dilution rate: it is even independent of the initial substrate concentration S_r. On the other hand, the biomass concentration depends on D and S_r. The relationships between x, *[S]*, the organism constants and D are demonstrated in Fig. 3.2.

In a batch culture the biomass concentration will increase rapidly and the maximal amount (as determined by the engineering parameters of the fermentor) will be reached within a few hours. If we start with a lower concentration of substrate and start to feed further substrate when it becomes exhausted, it is possible to approach the maximal amount of biomass in the fermentor slowly

(Fig. 3.3). Therefore, in a fed batch culture the capacity of the fermentor is utilized to a large extent for a much longer period than in a batch culture. This is main reason why the processes for industrial fermentations usually use fed batch cultures (Stouthamer and van Verseveld, 1987).

In addition to balance equations, kinetic ones are needed to describe microbial production processes. The linear equation for substrate consumption with product formation is the most important. The total rate at which substrate is used is the sum of the rates at which it is used for maintenance purposes, biomass and product formation (Stouthamer and van Verseveld, 1987).

$$r_s = m_s x_t + \frac{1}{Y_{xsm}} r_x + \frac{1}{Y_{psm}} r_p \qquad (3.31)$$

r_s, r_x and r_p are respectively the rates of substrate consumption, biomass and product formation and m_s and x_t are the maintenance coefficient and biomass concentration. Y_{xsm} and Y_{psm} are respectively the maximal yield of biomass and product on substrate.

The parameters of Eq. 3.31 are of great importance because they can be used:

- in mathematical models to optimize both the conversion of substrate into product, and the use of the bioreactor at full capacity;

- to analyze substrate-related production costs when a new process is being developed;

- to evaluate the yield of products of recombinant strains obtained through DNA technology along with the suitability of a host for producing a certain product.

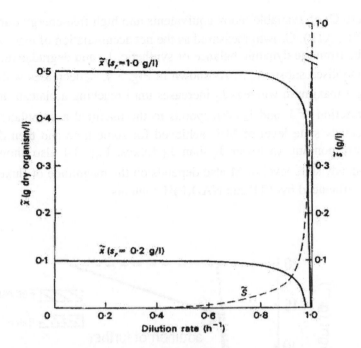

Figure 3.2. Steady state values of biomass (x) and growth-limiting substrate (s) concentrations in a chemostat. A plot of steady state values of x and s against dilution rate, D, with typical parameter values ($\mu_m = 1.0$ h^{-1}; Ks = 0.005 g/l; Y = 0.5) is shown (Reproduced from Pirt, S.J., 1975. *Principles of Microbe and Cell Cultivation.* ©copyright Blackwell Scientific Public).

Growth as a Balance of Fluxes

Any dynamic view of cell growth and proliferation, must consider growth as a dynamically coordinated, dissipative balance of fluxes, i.e. the result of fluxes of the different materials consumed, and the interactions between those fluxes (see also Chapter 5).

Growth as the result of the dynamic balance between synthesis and degradation of macromolecular components, e.g. DNA, RNA, proteins, polysaccharides, results in the typical growth curve shown by batch cultures of microorganisms. In this, synthetic process reactions decrease their rates with time, while degradative ones increase their rates, thereby achieving a pseudo-steady state level which will persist for some time and then decline (Aon and Cortassa, 1995). In Fig. 3.4 are shown the results of a numerical simulation of a typical growth curve of a batch culture of microorganisms at decreasing contents (from A to C) of available redox equivalents and high free-energy transfer bonds (NAD(P)H, ATP). Growth measured as the net accumulation of macromolecules, M, results from the dynamic balance of synthetic, J_S, and degradative, J_d, fluxes from catabolised substrate, S. As shown in Fig. 3.4, J_S decreases with time after reaching a maximum whereas J_d increases until reaching a plateau. In all cases, the intersection of J_s and J_d corresponds to the maximal accumulation of M. A pseudo-steady state level of M is achieved for some time and then decline; the latter corresponding to lower J_s than J_d fluxes. Fig. 3.4 also shows that the pseudo-steady state level of M also depends on the magnitude of fluxes which in turn are influenced by ATP and NAD(P)H contents.

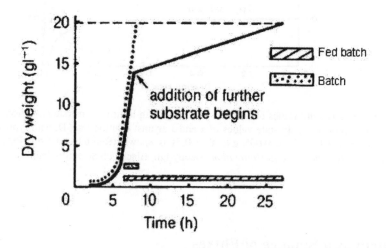

Figure 3.3. Growth of a microorganism in a batch and fed batch culture. At $t=0$ the culture was inoculated to a biomass concentration of 0.01 g l^{-1}. The division time of the organism was supposed to be 45 min. It was assumed that due to restricting engineering parameters the biomass concentration in the fermentor cannot exceed 20 g l^{-1}, since at higher concentrations the capacity for oxygen transfer of the fermentor is insufficient. Already after 8.1 h this limit is reached in the batch culture. If the substrate concentration is insufficient to reach this amount of biomass and a feed of substrate is started after the substrate has been fully consumed (indicated by an arrow) the limits of the fermentor capacity can be gradually approached. -----, biomass limit due to O$_2$ transfer; ·····, batch culture; ——, fed batch culture. Productive times in batch or fed batch cultures are indicated by horizontal bars. (Reprinted from *TIBTECH* 5, Stouthamer and van Verseveld. Microbial energetics should be considered in manipulating metabolism for biotechnological processes, 149-155. ©copyright 1987, with permission from Elsevier Science).

The analysis of the dynamics of cellular processes in the light of the concepts of dynamics and thermodynamics reveals that flux redirection at the cellular level may be accomplished taking advantage of the intrinsic dynamic properties of cellular metabolism (Aon and Cortassa, 1995, 1997) (see Chapter 5). Large dissipation rates through a metabolic path or branch flux redirection may take place because of kinetic limitations (see below). Kinetic limitation occurs when a metabolic pathway has attained the maximum possible rate of dissipation. The rate of dissipation suggests that at fixed ΔG_i then fixed stoichiometry, the maximum flux, J_i, may be set by the maximal through-put velocity sustained in that path (see Eq. 3.32 below). Since the maximum velocity depends on enzyme concentrations at rate-controling steps, then the higher the enzyme concentrations, the higher the flux.

The Flux Coordination Hypothesis

On the basis of the view of a cell as a dynamic balance of fluxes, we have introduced the Flux Coordination Hypothesis (FCH). This hypothesis emphasizes the regulation of the degree of coupling between catabolic and anabolic fluxes as a regulatory mechanism of the growth rate. FCH stresses the fact that cells exhibit global regulatory mechanisms of flux balance associated mainly with the rate of energy dissipation (Aon and Cortassa, 1997). Globally, the rate of energy dissipation by a cell is given by the product between fluxes, J_i, and their conjugated forces, ΔG_i, as follows:

$$\sigma = J_i \, \Delta G_i \qquad\qquad (3.32)$$

Taking into account the fact that the free-energy difference, ΔG_i, in every i-th enzyme-catalyzed reaction of either catabolism or anabolism is fixed by the nature of reactants and products, the effective flux to be sustained will depend on the genetic make up, as well as the availability of effectors according to environmental conditions (Aon and Cortassa, 1997). Thus, cells may redirect fluxes, modifying in this way the effective stoichiometry of metabolic pathways, specially when facing unfavorable environmental challenges (Fig. 3.5). Under these conditions cells uncouple catabolic and anabolic fluxes, becoming able to excrete products, some of them of commercial value. We may take advantage of these global regulatory mechanisms exhibited by cells for engineering them with the aim of obtaining specific products (see Chapter 7).

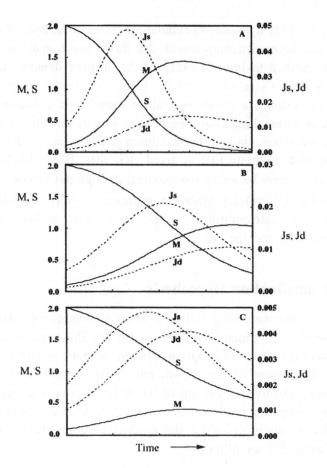

Figure 3.4. Cell growth as a result of a dynamic balance of synthetic and degradative fluxes. A simple model consisting of the autocatalytic synthesis of a macromolecular component (M) from a substrate (S). M is degraded according to a first-order kinetic law:

$$S + ATP + NADPH \rightarrow M \rightarrow$$

The model is mathematically expressed by two differential equations:

$$\frac{dM}{dt} = J_s - J_d$$

$$\frac{dS}{dt} = -J_s$$

with synthetic: $J_s = k_1 \cdot S \cdot ATP \cdot NADPH$; and degradative fluxes: $J_d = k_2 \cdot M$. k_1 and k_2 are the rate constants of synthesis and degradation of macromolecules (M), respectively. ATP and NADPH contents are parameters, i.e. they are assumed constant for simplicity. The parameter values were: $k_1 = 0.1$ (mM^{-2} h^{-1}); $k_2 = 0.01$ (h^{-1}); S_0 (the initial concentration of substrate=2.0; $ATP = 1.0$. According to the units employed in the model, S, ATP, NADPH and M are in mM concentration units. From A to C the varying parameter is NADPH: (A) 0.5; (B) 0.25; (C) 0.1; indicating decreasing redox equivalent contents. Notice that when the flux of synthesis (J_s) intercepts the flux of degradation

(J_d) the accumulation of M is maximal; thus, when $J_s > J_d$ the flux balance favors the synthesis of M and consequently it accumulates. On the contrary, when $J_s < J_d$ the flux balance favors the degradation of M which begins to decline. The time as well the amount reached by M depends on the metabolic status, e.g. available redox equivalents (Reprinted from *Prog. Biophys. Mol. Biol.* 64, Aon and Cortassa. Cell growth and differentiation from the perspective of dynamical organization of cellular and subcellular processes, 55-79. ©copyright 1995, with permission from Elsevier Science).

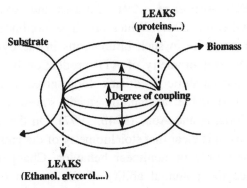

Figure 3.5. The Flux Coordination Hypothesis. The degree of coupling between catabolic and anabolic fluxes as a phenomenon involved in the regulation of microbial proliferation and metabolic flux redirection. The hypothesis postulates that for higher degrees of coupling, i.e. longest arrow, the more coupled anabolic and catabolic fluxes will be, and that would correspond to lower leaks (dashed arrows). Under those conditions, cells will continue their mitotic cycling. Cells will leave the mitotic cycle whenever they are challenged by an unfavorable environment, or when environmental changes, e.g. oxygen, carbon source, induce a particular metabolic and energetic status that lead to a differential gene expression which in turn induce metabolic flux redirection and lower growth rates (Reprinted from *Prog. Biophys. Mol. Biol.* 64, Aon and Cortassa. Cell growth and differentiation from the perspective of dynamical organization of cellular and subcellular processes, 55-79. ©copyright 1995, with permission from Elsevier Science).

An imbalance between catabolic and anabolic fluxes may be associated with the onset of cell cycle arrest through growth limitation (Fig. 3.5). According to FCH, some cellular mechanism for metabolic fluxes redirection toward product (either catabolic, organic acids, or anabolic, polysaccharides) or biomass (macromolecules) formation should be associated with the onset of cell division arrest (see Chapters 4 and 7). This is indeed very relevant since through regulation of the growth rate, an effective change is attained in the length of the cell cycle, and thus of gene expression. In this way, the cell cycle becomes a potential (supra)regulatory mechanism of gene expression (Gubb, 1993; Aon and Cortassa, 1997). We have shown that, independently of the way the growth rate was varied, the time consumed by a yeast cell to stay in the G1 phase of the cell cycle, decreases exponentially with growth rate (Fig. 3.6; see also Chapter 7). This is in agreement with the fact that the G1 phase of the cell cycle is the most sensitive to environmental conditions, and is coincident with the main molecular components acting at Start (Hartwell, 1991; Forsburg and Nurse, 1991; Reed, 1992; Aon and Cortassa, 1995, 1998).

Flux redirection may be viewed as a bifurcation in the phase portrait dynamics of metabolic flux behavior (Abraham and Shaw, 1987; Bailey *et al.*, 1987; Kauffman, 1989; Aon *et al.*, 1991; Cortassa and Aon, 1994b; Aon and Cortassa, 1997; Lloyd *et al.*, 2001). Which are those nonlinear mechanisms that at the cellular level give rise to such bifurcations in the dynamic behavior of metabolic fluxes? The stoichiometry of ATP production by glycolysis is itself autocatalytic. This autocatalytic feed-back loop gives rise to sophisticated dynamics in addition to stable, asymptotic steady states, and oscillatory ones (Aon *et al.*, 1991; Cortassa and Aon, 1994b, 1997). On the basis of the results obtained for glycolysis, it is clear that stoichiometry of the pathways constitutes a built-in autocatalytic source of nonlinear behavior. Changes in stoichiometry would then introduce the potential ability of the system to exhibit different dynamics in metabolic fluxes.

Toward a Rational Design of Cells

Our aim here is to debate whether there is a rational and quantitative approach to the modification of a microorganism for a specific biotransformation process, or for the production of high-value chemicals.

In the following section we explain the main steps to be taken in order to proceed through the flow diagram corresponding to the TDA approach described in Chapter 1 (Fig. 1.2). For our purpose, we have at hand two main methodologies: Metabolic Flux Analysis (MFA) and Metabolic Control Analysis (MCA). By applying the TDA approach (Aon and Cortassa, 1997; Aon and Cortassa, 1998; Cortassa and Aon, 1997, 1998) we start to analyze the natural potential of the microorganism for production of the chemical of interest (see Chapter 4). It is preferable to accomplish this first step of the TDA approach in chemostat cultures; this allows us to detect the phase of most interest for the production of the metabolite. For instance, if the metabolite production is sensitive to the growth rate, it may be studied as a function of the dilution rate, D (D=growth rate, μ). Under these conditions, MFA or other methods may be applied in order to quantify the yield of the metabolite under the culture conditions employed. We must compare the yield obtained with the theoretical yield, in order to decide whether performing genetic engineering to obtain a recombinant strain is worthwhile.

Several criteria may be adopted when deciding the feasibility of modifying a cell or microorganism through DNA recombinant technology: *(i)* how far are we below the theoretical yield; *(ii)* if we are close to the theoretical yield (say 90-95%), then we must consider whether a 5% to 10% increase in the production of the desired metabolite is economically worthwhile. This decision depends on the commercial value of the product and its concentration before downstream processing. The higher the starting concentration the less costly is the final product (Fig. 3.7).

Once it has been decided to genetically modify the organism, a strategy must be designed. The first step of this strategy is to decide which steps of the metabolic pathway are appropriate.

In the next sections we will show different approaches by which MFA will help to design and optimize a strategy for the genetic modification of a microorganism for a specific biotechnological aim. A more extensive and detailed treatment is performed in Chapter 4.

Redirecting Central Metabolic Pathways under Kinetic or Thermodynamic Control

In metabolism, central catabolic pathways (glycolysis, pentose phosphate pathway, tricarboxylic acid cycle, oxidative phosphorylation) provide the key intermediary metabolites precursors of monomer synthesis (amino acids, lipids, sugars, nucleotides) for the main macromolecular components (proteins, lipids, polysaccharides, nucleic acids) (Fig. 2.2). The production rate and yield of a metabolite are ultimately limited by the ability of cells to channel the carbon flux from central catabolic pathways to main anabolic routes leading to biomass synthesis (Cortassa *et al.*, 1995; Liao *et al.*, 1996; Vaseghi *et al.*, 1999). Thus, an important question in MCE is how the fluxes towards key precursors of cell biomass are controlled. This is a very important topic beyond the peculiarities established by the production of a desired metabolite or molecule since it points out toward general constraints of biological performance.

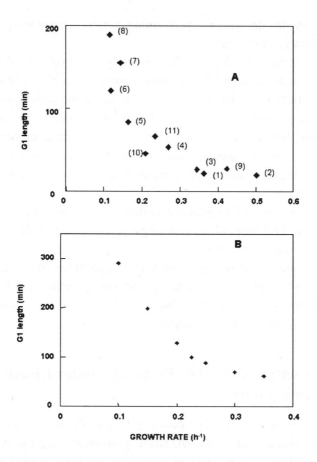

Figure 3.6. G1 duration as a function of the growth rate shown by the WT CEN.PK122 and catabolite (de)repression mutants *snf1*, *snf4* and *mig1* in batch cultures (A) and aerobic, glucose-limited, chemostat cultures (B). (A) The values of G1 lengths presented correspond to logarithmically growing batch cultures of the strains on 4% glucose, 3% ethanol or 4% glycerol as described in Aon and Cortassa (1999). The numbers correspond to: *mig1*, 4% glucose, synchronized cultures with mating pheromone (1); WT, 4% glucose, synchronized cultures with mating pheromone (2); WT, 4% glucose (3); *mig1*, 4% glucose (4); *mig1*, 3% ethanol (5); WT, 3% ethanol (6); *mig1*, 4% glycerol (7); WT, 4% glycerol (8); *snf4*, 4% glucose, synchronized cultures with mating pheromone (9); *snf4*, 4% glucose (10); *snf1*, 4% glucose (11). In (B) are represented the data of G1 length obtained from aerobic, glucose-limited chemostat cultures run under the conditions described in Cortassa and Aon (1998) and Aon and Cortassa (1998). After attaining the steady state, judged through the constancy in biomass concentration in the reactor vessel and in the concentration of O_2 and CO_2 in the exhaust gas, samples were taken for staining with propidium iodide and subsequent flow cytometry analysis (see Aon and Cortassa, 1998). The growth rate shown in panel (B) is fixed by the experimenter through the dilution rate, D, and not by the strain or the carbon source as shown in panel (A). (Reproduced from *Current Microbiology*. Quantitation of the effects of disruption of catabolite (de)repression genes on the cell cycle behavior of *Saccharomyces cerevisiae*. Aon and Cortassa, 38, 57-60, Fig. 2, 1999. ©copyright Springer-Verlag, with permission)

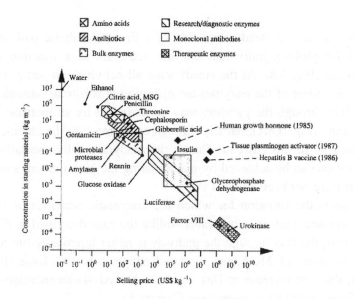

Figure 3.7. Relationship between selling price and concentration before downstream processing for several fermentation products. (Reproduced from Doran, P.M. *Bioprocess Engineering Principles*. ©copyright 1995 Academic Press, with permission)

A large experience in the physiological behavior of microorganisms shows that their ability to channel carbon flux from central catabolic pathways to main anabolic routes is either thermodynamically- or kinetically-controlled. This distinction is timely topic when the performance of a microorganism is aimed to be improved through MCE or optimization of environmental conditions. In the following we focus to show that distinguishing about kinetic or thermodynamic control of metabolic flux under a particular growth condition in which a cell (e.g. a unicellular microorganism) is searched to be improved, is critical for two main reasons: *(i)* the decision about the strategy of engineering by acting on metabolism or optimization of culture conditions; and *(ii)* the evaluation of cellular performance, either genetically engineered or not, under a defined condition.

Thermodynamic or Kinetic Control of Flux under Steady State Conditions

In terms of enzyme kinetics, a steady state flux of a linear metabolic pathway may be visualized by plotting individual enzyme activities as a function of substrate concentration (Fig. 3.8). At the steady state all activities are set at the same value. In the case none of the enzymes are operating at substrate saturation (i.e. at *Vmax*), the flux through the pathway may increase just by an increase in the amount of substrate (Fig. 3.8A), namely through an increase in the chemical potential of the reaction. In this case, the flux is thermodynamically-controlled, and further optimization can be achieved by manipulating the environment since the biological limits have not been attained yet.

Figure 3.8B depicts the situation for which the enzymatic activities of the pathway are operating near their maximal rates, unlike the case described in Fig. 3.8A. At plateau levels, the flux through the pathway is rather kinetically-limited than by substrate. In terms of MCA, the saturated enzymes have a large flux control coefficient; thus, an increase in flux can be achieved via an increase in the amount of the flux-controling enzymes (see Chapter 4).

An analogy may further help to clarify thermodynamic or kinetic control. The metabolite flow being sustained through metabolic pathways in a cell would be analogous to a river's basin whose depth determines the extent of the maximal flux that the river is able to sustain. Under thermodynamic limitation the basin is large with respect to the water provision; thus, the cell is able to sustain even larger fluxes. Under those conditions, the process can be further optimized by manipulating the environment, as the biological limits have not been attained. On the other hand, in the case of kinetic limitation, the improvement of the flux will require manipulation of the metabolic pathway through genetic engineering.

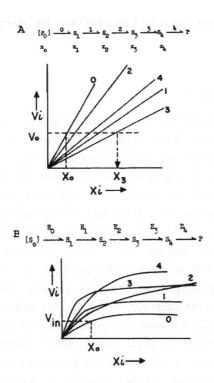

Figure 3.8. Initial rate of a hypothetical metabolic pathway as a function of substrate concentration. Each enzyme-catalyzed step of the linear metabolic pathway at the top of each panel exhibits a dependence on substrate concentration, X, as indicated in the diagrams. (A) All enzymes display linear kinetics in the physiological range of variation; at the steady state the rate of each enzyme is equal to V_O and the level of each substrate is given by the intersection between enzyme rate, Vi, and the V_O line. (B) The enzymes of the linear pathway operate near saturation and the enzyme activity E_O corresponds to the rate-controling step. As in panel A, the steady state flux is set at the V_{in} value as well as all enzyme rates (i.e. steady state condition); the substrate concentrations attained are given by the intersection between the corresponding rate curve and the V_{in} line. (Adapted from Higgins, J.J. *Control of Energy Metabolism*, pp. 13-46. ©copyright 1965 Academic Press, with permission).

The point at which the thermodynamic limitation is achieved can be judged through the yield which is in fact, a thermodynamic parameter (Heijnen and van Dijken, 1992; Roels, 1983). The latter points out the importance of a correct evaluation of biomass yields during microbial growth (Cortassa *et al.*, 1995; Verduyn, 1992).

Kinetic and Thermodynamic Limitations in Microbial Systems. Case Studies

Saccharomyces cerevisiae

As an example let us discuss the case of the yeast *S. cerevisiae* growing in aerobic, glucose- or nitrogen-limited chemostat cultures. The maximal theoretical anabolic flux necessary for yeast cells growing in the respiratory regime of continuous cultures can be calculated. It has been found that during solely respiratory glucose breakdown, the flux through anabolic pathways (i.e. that directed toward the synthesis of cell biomass) is thermodynamically-controlled (Aon, J.C. and Cortassa, 2001). This means that if we add more substrate, the anabolic flux could still increase, or in other words the pools of the key precursors (e.g. hexose-6-P, triose-3-P, ribose-5-P, erythrose-4-P, 3-P-glycerate) of macromolecules (lipids, proteins, polysaccharides, nucleic acids) are not saturated with respect to the maximal flux attainable under such conditions. On the contrary, in the so called respiro-fermentative regime of yeast physiology in chemostat cultures, the fluxes through central anabolic pathways are kinetically-controlled. The accumulation of substrate (e.g. glucose) and of catabolic by-products (e.g. ethanol, glycerol) in the extracellular medium, provides evidence for anabolic flux saturation. Stated otherwise, the amount of intermediates necessary for growth are far in excess of theoretical needs. Under these conditions, catabolism uncouples from anabolism, and catabolic or anabolic products start to accumulate either intra- or extra-cellularly.

In the thermodynamically-limited regime (i.e. at low D), the biomass appeared to be more sensitive to N-limitation (i.e. an anabolic limitation) than at high D where the limitation was mainly kinetic. Figure 3.9 shows the carbon fate observed in the respiratory regime or in the respiro-fermentative mode of glucose breakdown either in ammonia- or glutamate-fed cultures, for which the carbon recovery was complete (Aon, J.C. and Cortassa, 2001). Most carbon was recovered as cell biomass under C-limitation in the respiratory regime either with glucose (54%) or glucose plus glutamate (63%) (Fig. 3.9 A,E). On the contrary, carbon mostly evolved as CO_2 under N-limitation; this trend being more drastic in the presence of mixed carbon substrates (glucose plus glutamate) (Fig. 3.9, compare pies C and G). Under N-limited conditions, acetate and glycerol were excreted along with some ethanol (Fig. 3.9), although q_{EtOH} was low enough (< 1.0 mmol h^{-1} g^{-1} dw) to be considered as a respiratory regime (Aon, J.C. and Cortassa, 2001). In the presence of ammonia, during the respiro-fermentative regime, (i.e. high D), large differences in carbon distribution were observed between C- and N-limited cultures, especially with glucose as the C-source. In fact, cell biomass decreased from 34% to 16 % under N-limitation with respect to C-limitation with a two-fold increase in ethanol recovery, i.e. from 27% to 52% (Fig. 3.9 B,D). Concomitantly, carbon redistributed toward acetate and glycerol (3%), and CO_2 decreased from 39 to 29%. Interestingly, the fate of carbon was similar under C- or N-limitation despite cells were performing a mixed-substrate utilization (glucose plus glutamate) (Fig. 3.9 F,H). Apparently, glutamate was able to provide precursors for biomass synthesis allowing more glucose to be directed toward ethanolic fermentation. However, the similar carbon distributions were obtained from very different patterns of metabolic flux (Aon, J.C. and Cortassa, 2001). A quite different picture was observed in N-limited cultures, both during purely respiratory and respiro-fermentative glucose breakdown modes (Aon, J.C. and Cortassa, 2001). From the total glucose consumption flux, only 26% was directed to biomass synthesis at Dc whereas at high D this percentage decreased to only 15% of the total flux. Due to the excess carbon present in N-limited cultures, the contribution of glutamate to biomass precursors was less important and limited to the provision of carbon through α ketoglutarate- and oxalacetate-derived C compounds (Aon, J.C. and Cortassa, 2001).

A threshold value of glucose consumption rate, q_{Glc}, was determined at the critical dilution rate, Dc, just before the onset of the respiro-fermentative regime (Aon, J.C. and Cortassa, 2001). In fact, the glucose consumption rate at Dc was the same, independent of the nature of the nutrient limitation, i.e. carbon or nitrogen, and only slightly dependent upon the quality of the N-source. The latter may be indicating a threshold for glucose consumption at which yeast cells are no longer able to process glucose solely through the oxidative pathway. It also denotes the achievement of a biological limit upon which a redirection of the catabolic carbon flux through ethanolic fermentation, is triggered.

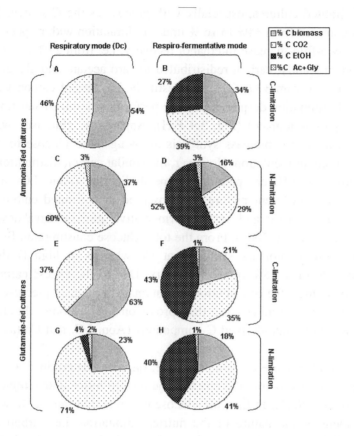

Figure 3.9. Steady state carbon distribution under C- or N-limitation in the respiratory or respiro-fermentative regimes of glucose breakdown. The pies indicate the fate of carbon from glucose (panels A-D) or from glucose plus glutamate (panels E-H) at the critical dilution rate, Dc (panels A, C, E, G, respiratory metabolism), and at a high dilution rate, D (0.34-0.36 h^{-1}) (panels B, D, F, H, respiro-fermentative metabolism) from chemostat cultures run under C-limitation (A,B,E,F) or under N-limited conditions (C,D,G,H). In panels A-D, the N-source was ammonia, while glutamate was the N-source in panels E-H. When the percentage carbon recovered as a given compound was 1% or over, the value is indicated near its "portion" in the pie (Reproduced from *Metabolic Engineering*, Aon, J.C. and Cortassa, ©copyright 2001 Academic Press, with permission).

The interaction between carbon and nitrogen metabolism set the level of anabolic flux since both, C- and N-sources supply intermediates for biosynthesis. The role of the nitrogen metabolism in the triggering of ethanol production would be realized through setting the anabolic flux and in turn the biomass level in the chemostats. These biomass levels determine when the threshold glucose consumption rate is achieved after which ethanolic fermentation is triggered.

Escherichia coli

This microorganism was selected for exploiting rich medium for rapid growth as this organism can sustain very high maximal rates of gene expression of the components of the machinery for protein and RNA synthesis (Jensen and Pedersen, 1990). In this growth mode, biosynthetic mechanisms are highly dependent on: *(i)* the accumulation of large pools of activated precursors (e.g. charged tRNAs complexed with GTP and elongation factor (EF)-Tu), and *(ii)* free catalytic components (ribosomes). Upon achievement of maximal rates of biosynthesis, *E. coli* becomes depleted of precursors for macromolecules synthesis, as the cell's ability to use the activated precursors and catalytic components exceeds the capacity of intermediary metabolism to provide these precursors (Jensen and Pedersen, 1990). By analogy with the growth condition in a chemostat, the bacterium is intrinsically able to grow faster than it is allowed to by the peristaltic pump feeding the vessel. This is a similar case as the one described above for yeast growing in chemostat culture in the respiratory regime (i.e. low growth rates). We stated that under those conditions the anabolic flux is thermodynamically-controlled (in the sense described above, i.e. the Gibbs free energy contributed by the substrate is limiting growth), and are not saturated with respect to the key intermediary precursors of the biomass.

Jensen and Pedersen (1990) pointed out that the transcription and translation steps, are the ones involved in the control of growth rate and macromolecular composition. Nevertheless, this view does not hold for very high growth rates in which the situation becomes the inverse of the one described. This will be the case for yeast growing at a high growth rate in chemostat culture where the pools of precursors are saturated, and the system is kinetically-controlled. Apparently, the same happens for *E. coli* grown in glucose-limited chemostat cultures, i.e. by-product secretion occurs at a growth rate of 0.9 h^{-1}, at a maximal oxygen uptake rate of 20 mmol O_2 g^{-1} dry weight h^{-1} (Varma *et al.*, 1993) (Fig. 3.10). Acetate is the first by-product secreted as oxygen becomes limiting, followed by formate, and then by ethanol at even higher growth rates (Varma *et al.*, 1993). *E. coli* excretes 10-30% of carbon flux from glucose as acetate under aerobiosis (Holms, 1986).

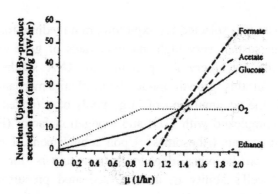

Figure 3.10. Optimal aerobic growth and secretion of by-products predicted by a flux balance model for *E. coli* during glucose limited growth. (Reprinted with permission from Nature. Varma and Palsson, *Bio/Technology* 12, 994-998. ©copyright 1994 Macmillan Magazines Limited).

Flux redirection may also arise when either phosphorylation or redox potentials become unbalanced, e.g. due to restriction of electron acceptors in aerobic bacteria (Verdoni *et al.*, 1990, 1992).

Increasing Carbon Flow to Aromatic Biosynthesis in *Escherichia coli*

Based on the stoichiometric models of metabolism that constitute the basis of MFA, the first step is to identify possible pathways leading to the production of the metabolite(s) of interest. The main aim of this analysis is then to identify all stoichiometrically plausible pathways, determine flux distributions, and calculate theoretical yields.

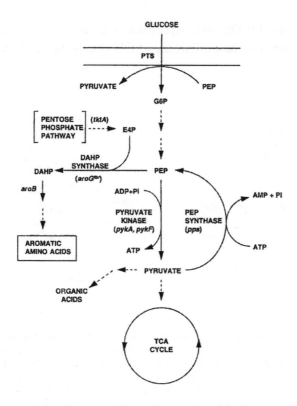

Figure 3.11. Metabolic pathways related to the formation and consumption of PEP, and the relationship of PEP to aromatic amino acid biosynthesis. Dashed lines represent multiple enzymatic steps. (Reprinted with permission from Nature Publishing Group. Gosset, Yong-Xiao and Berry, *J. Ind. Microbiol.* 17, 47-52. ©copyright 1996).

Figure 3.11 shows the central metabolic pathways related to the formation and consumption of PEP, and the pathways leading to aromatic amino acid biosynthesis (Gosset *et al.*, 1996). As can be seen in Fig. 3.11, PEP is used both in the aromatic biosynthesis and in the transport of glucose via a phosphotransferase system (PTS). To increase the intracellular levels of PEP may therefore help to divert the carbon flux to aromatic amino acid synthesis. After PEP is converted to pyruvate during glucose transport, pyruvate is not recycled to PEP under glycolytic conditions. Pyruvate recycling to PEP, or non-PTS sugars (e.g. xylose), may be used to avoid wastage of pyruvate (Liao *et al.*, 1996). Another possibility for increasing the intracellular concentration of PEP is to inactivate one or both genes coding for pyruvate kinase in *E. coli* (*pykA, pykF*)

(Gosset *et al.*, 1996). Finally, overexpression of transketolase coded by the *tktA* gene, in order to augment the availability of E4P, has been also tested.

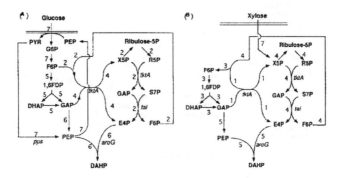

Figure 3.12. Optimal pathways and flux distribution involved in 3-deoxy-D-arabino-heptulosonate-7-phosphate (DHAP) synthesis from glucose (A) or xylose (B). The pathways that convert sugars to aromatic metabolites in *E. coli* has been examined by Liao *et al.* (1996). The first step after commitment to the aromatic pathway is the condensation between PEP and erythrose 4-phosphate (E4P) to form 3-deoxy-D-arabino-heptulosonate-7-phosphate (DHAP), that is catalyzed by DAHP synthase. To produce aromatic metabolites from glucose, carbon flow has to be effectively channelled through DAHP synthase. Typically, the carbon yield of aromatic metabolites from glucose is less than 30%. (Pathways analysis, engineering, and physiological considerations for redirecting central metabolism. Liao, Hou and Chao, *Biotechnol. Bioengin.* 52, 129-140. ©copyright 1996 John Wiley & Sons, Inc. Reprinted by permission of Wiley-Liss, Inc., a subsidiary of John Wiley & Sons, Inc.)

The strain choice is an important step in the design of a strategy to obtain recombinants. Since the condensation of E4P and PEP to form DAHP is irreversible, and since DAHP has no other known function in the cell, accumulation of DAHP is a good indicator of carbon commitment to aromatic biosynthesis. An *aroB* mutant of *E. coli* that cannot metabolise DAHP further and excretes this metabolite to the medium has been used (Liao *et al.*, 1996; Gosset *et al.*, 1996). Even wild-type *E. coli* strains with a functional *aroB* gene and modified with a plasmid which overproduces DAHP, could excrete the precursor into the medium (Gosset *et al.*, 1996).

Table 3.3. Increasing carbon flow to aromatic biosynthesis in *Escherichia coli*. DAHP production and doubling times.

Strain	Relevant property		mmol DAH(P) g^{-1} dry cell weight		Doubling time (h)	
			Mean[a]	s.d.[b]	Mean	s.d.
1	PB 103(pRW300)	Control	0.37	0.08	1.91	0.07
2	PB 103A(pRW300)	*pykA*⁻	0.54	0.16	1.85	0.11
3	PB103F(pP\W300)	*pykF*⁻	0.44	0.07	1.83	0.12
4	PB 103AF(pRW300)	*pykA*⁻ *pykF*⁻	1.24	0.08	1.91	0.11
5	PB103(pRW300, pCLtkt)	*tktA*⁺⁺	1.66	0.26	1.78	0.24
6	PB103A(pRW300, pCLtkt)	*pykA*⁻ *tktA*⁺⁺	1.81	0.20	1.75	0.21
7	PB103F(pRW300, pCLtkt)	*pykF*⁻ *tktA*⁺⁺	1.63	0.05	1.76	0.22
8	PB 103AF(pRW300, pCLtkt)	*pykA*⁻ *pykF*⁻ *tktA*⁺⁺	1.94	0.12	1.85	0.07
9	PB 103(pRW5)	*Control*	0.77	0.07	2.07	0.37
10	PB 103(pRW5tkt)	*tktA*⁺⁺	1.65	0.23	2.06	0.46
11	PB103(pRW5, pPS341)	*pps*⁺⁺	1.37	0.30	2.05	0.40
12	PB 103(pRW5tkt, pPS341)	*pps*⁺⁺ *tktA*⁺⁺	2.93	0.63	2.15	0.59
13	NF9(pRW300)	*PTS*⁻	0.58	0.24	1.88	0.12
14	NF9A(pRW300)	*PTS*⁻ *pykA*⁻	1.61	0.68	2.19	0.52
15	NF9F(pRW300)	*PTS*⁻ *pykF*⁻	0.26	0.12	1.95	0.33
16	NF9AF(pRW300)	*PTS*⁻ *pykA*⁻ *pykF*⁻	3.39	0.68	2.26	0.35
17	NF9(pRW300, pCLtkt)	*PTS*⁻ *tktA*⁺⁺	2.15	0.94	2.00	0.14
18	NF9A(pRW300, pCLtkt)	*PTS*⁻ *pykA*⁻ *tktA*⁺⁺	3.29	0.59	1.85	0.26
19	NF9F(pRW300. pCLtkt)	*PTS*⁻ *pykF*⁻ *tktA*⁺⁺	1.18	0.60	1.87	0.13
20	NF9AF(pRW300, pCLtkt)	*PTS*⁻ *pykA*⁻ *pykF*⁻ *tktA*⁺⁺	7.37	0.37	3.24	0.37

[a] Results shown are averages of three independent experiments. Within each experiment, strains were tested in duplicate or triplicate.
[b] Standard deviation.
(Reprinted with permission from Nature Publishing Group. Gosset, Yong-Xiao and Berry, *J. Ind. Microbiol.* 17, 47-52. ©copyright 1996).

The recycling of pyruvate to PEP can be achieved by overexpression of PEP synthase (*Pps*) in the presence of glucose. The optimal flux distributions for the strategies in which pyruvate is recycled or non-PTS sugars were used, are shown in Fig. 3.12. These approaches all allow 100% theoretical carbon yield, with 86% theoretical yield of DAHP from hexose, or 71% theoretical molar yield when non-PTS sugars are used, even without recycling pyruvate to PEP (Liao *et al.*, 1996).

In Table 3.3 it can be seen that the overexpression of *tktA* produces a significant increase in the flux directed to DAHP synthesis (Gosset *et al.*, 1996). Within the framework of MCA, it may be deduced that transketolase represents a rate-controling step of the carbon flux committed to aromatic amino acid synthesis. The latter increase was verified both with control and PTS⁻ glucose⁺ strains (Gosset *et al.*, 1996; Flores *et al.*, 1996). Some improvement could also be obtained with Pps overexpression. The highest yield of DAHP was achieved with strains in a PTS-background, in which *pykA* and *pykF* were deleted, and *tktA* overexpressed. This latter result is another illustration of a shared control of the flux by several participating steps in aromatic amino acid synthesis.

The significance of each gene in the production of aromatics was also evaluated by Liao *et al.* (1996) in terms of the DAHP yield from the carbon source, either glucose or xylose. The host, an *aroB* strain that cannot metabolise DAHP further and excretes this metabolite to the medium, was found to produce very little DAHP without the overexpression of *aroG*. With *aroG* overexpression alone, this strain gave about 60% molar yield of DAHP from glucose (Liao *et al.*, 1996). This yield was greater than the theoretical one (43%), without recycling pyruvate to PEP.

Another pathway had to be invoked for pyruvate recycling since the yield obtained was higher than the expected one. The pyruvate recycling through the basal level of Pps could not explain the yield of 60% obtained, because knocking out the chromosomal *pps* gene showed no effect on the DAHP yield. It was found that pyruvate recycling can be also mediated through the glyoxylate shunt and the Pck reaction; the optimal yield for this pathway being 64% (Liao *et al.*, 1996). When both *Tkt* and *Pps* were overexpressed along with *AroG*, the molar yield was as high as 94%.

Chapter 4

Methods of Quantitation of Cellular "Processes Performance"

Stoichiometry of Growth: The Equivalence between Biochemical Stoichiometries and Physiological Parameters

All methods of Metabolic Flux Analysis (MFA) begin with the evaluation of the rate of change of concentration of a metabolic intermediate. This may be estimated in general terms as the sum of all fluxes producing the metabolite, minus the sum of all fluxes consuming it; each one of the fluxes being multiplied by the corresponding stoichiometric coefficient. The flux is a global, systemic property, resulting from the rate of conversion of metabolic precursors into products within metabolic pathways. Two techniques are primarily used for flux determination: mass isotopomer analysis and extracellular metabolite balance models (Yarmush and Berthiaume, 1997; Fiehn *et al.*, 2000; Stephanopoulos, 2000; Raamsdonk *et al.*, 2001). Several technologies are used to perform mass isotopomer analysis such as gas chromatography/mass spectrometry (GC/MS) and nuclear magnetic resonance (NMR) spectroscopy. GC/MS technology is considered one of the mature technologies available at present for assessing the metabolite profiling of cells (Fiehn *et al.*, 2000).

In order to calculate the flux through anabolic pathways for biomass synthesis, stoichiometries together with the amount of precursors required for biomass should be multiplied by the growth rate. The latter is a constant equal to the dilution rate at the steady state regime in chemostat cultures (see Chapters 3 and 5). In batch cultures, it is directly the rate of increase in cellular dry weight if a condition of balanced growth can be demonstrated (Cortassa *et al.*, 1995). The demand of precursors for biomass synthesis is calculated from the biomass composition determined under each growth condition. Thus, to apply MFA to the estimation of anabolic fluxes, apart from the knowledge of the metabolic pathways and their stoichiometries, two more pieces of information are required: the biomass composition and the growth rate.

Table 4.1. Metabolic fluxes and the equivalence between biochemical stoichiometries and physiological parameters for yeast cells growing on different carbon sources in minimal medium

Carbon Source	Growth rate[a] (1/h)	P/O ratio	Fluxes[b]				Y_{ATP}^{max}[c]	theo Y_S[d]	exp Y_S[e]	Yield index[f]
			PP pathway	oxidative catabolic	anabolic	total carbon				
Glucose	0.307	g	0.049	0.90	2.2	3.2	24	95	28 ± 3	0.30
	± 0.005	1	0.10	0.59		2.9	36	103		0.27
		2	0.17	0.18		2.6	76	117		0.24
Glycerol	0.18	g	0.0067	1.2	2.5	3.7	19	47	15 ± 2	0.33
	± 0.02	1	0.11	0.57		3.2	39	55		0.28
Pyruvate	0.15	g	0.00	2.2	1.2	4.5	13	33	10 ± 2	0.32
	± 0.02	1	0.00	2.1		4.4	14	33		0.31
		2	0.12	1.3		3.7	14	40		0.26
		3	0.20	0.85		3.3	15	44		0.23
Lactate	0.156	g	0.081	1.9	2.5	4.5	13	34	31 ± 4	0.92
	± 0.004	1	0.099	1.8		4.5	14	35		0.90
		2	0.23	1.1		3.8	15	41		0.77
		3	0.29	0.68		3.5	15	45		0.71
Ethanol	0.205	g	0.00	5.7	5.2	11	8.9	19	31 ± 2	>1
	±0.002	1	0.00	3.6		8.8	14	23		>1
		2	0.00	1.1		6.4	26	32		0.84
Acetate	0.16	g	0.00	6.3	4.2	10	8.7	16	17 ± 3	>1
	±0.01	1	0.00	5.5		9.6	10	17		>1
		2	0.00	3.1		7.3	11	23		0.77
		3	0.00	1.6		5.8	13	28		0.62

[a] Experimental growth rate (μ) ± SEM used for calculation of the flux through the PP pathway, oxidative catabolism and the total (catabolic + anabolic) flux of substrate. It is equivalent to ln 2/(doubling time).

[b] Expressed as mmol carbon substrate h^{-1} g^{-1} dw.

[c] Y_{ATP}^{max} is the theoretical yield calculated from the known requirements of ATP for synthesis, polymerization and transport (see Materials and Methods) expressed as g yeast dry weight per mol ATP.

[d] molar growth yield defined as the amount of biomass (g yeast dry weight) synthesized from one mol of carbon substrate.

[e] experimentally determined molar growth yield for CH1211 strain of *S. cerevisiae* growing in minimal medium on different carbon substrates, expressed as g dw mol^{-1} carbon substrate ± SEM.

[f] yield index defined as the ratio of the theoretical substrate flux required to sustain the experimental growth rate to the experimental flux of substrate consumption. It may also be obtained from the ratio exptl Y_S (column 10) over theor Y_S (column 9) since the latter are inversely proportional to the substrate consumption rates.

[g] calculation performed without taking into account the oxidation of NADH produced in anabolism

by the mitochondrial electron transport chain.
(Fluxes of carbon, phosphorylation, and redox intermediates during growth of *Saccharomyces cerevisiae* on different carbon sources. Cortassa, Aon and Aon, *Biotechnol. Bioengin.* 47, 193-208. ©copyright 1995 John Wiley & Sons, Inc. Reprinted by permission of Wiley-Liss, Inc., a subsidiary of John Wiley & Sons, Inc.).

When an insufficient number of fluxes are experimentally measurable, MFA may be performed on the basis of optimization principles (see below). In Cortassa *et al.* (1995) the optimization principle based on a minimal energy dissipation, was applied; this assumes that the function of the tricarboxylic acid (TCA) cycle and the respiratory chain is to generate enough ATP, and of the pentose phosphate (PP) pathway to produce adequate amounts of reducing equivalents to fulfil biosynthetic demands. The addition of the carbon flux required to provide carbon intermediates (macromolecular precursors) to the catabolic flux (TCA cycle and PP pathways) gives the minimal flux of substrate consumption per gram of biomass dry weight. The reciprocal of this minimal flux multiplied by the growth rate gives the maximal yield of biomass on carbon substrate (Table 4.1). Following a similar procedure, the minimal catabolic flux of ATP required to accomplish the biosynthetic energy demand, gives the flux of ATP formation. The maximal yield on ATP is given by the quotient of the growth rate to ATP flux.

Both yields, Y_{SX} and Y_{ATP}, are highly dependent on the P:O ratio (i.e. the mole numbers of ATP synthesized per mole of electron pairs transported by the respiratory chain). This ratio gives a measure of the efficiency of oxidative phosphorylation (see Chapter 2 Section: *Catabolic fluxes*). As the criterion for calculating the catabolic flux is to minimize the ATP flux, the efficiency of oxidative phosphorylation will have a large influence on such a calculation. Furthermore, if the redox potential generated during synthesis of monomer precursors of macromolecules is considered to be reoxidised in the respiratory chain, thereby contributing to the generation of ATP, this will also influence maximal values of Y_{SX} and Y_{ATP}. As an example, Table 4.1 shows both yields as a function of different P:O ratios. If as a function of a given P:O ratio, the ATP flux gave a negative figure, this value was omitted (e.g. P:O=3 in glucose, P:O=2 or 3 in glycerol) (see Cortassa *et al.*, 1995). An increase in the efficiency of oxidative phosphorylation will result in a lower catabolic flux since the same amount of ATP will be generated by a comparatively lower amount of substrate. This decrease may become very significant in yeast cells growing on glycerol at P:O larger than 1, or glucose or ethanol for P:O=3. If all the NADH produced in anabolic Steps I and II were reoxidized to produce ATP, this would largely overwhelm the requirement of phosphorylation energy for biosynthesis. Under those conditions, no extra carbon would be required to be catabolised to fulfil anabolic demand of high-energy transfer phosphate bonds (~P) (Table 4.1). The catabolic utilization of the PP pathway in a cyclical-functioning mode (i.e. generation of NADPH with the concomitant oxidation of the substrate to CO_2) was altered depending on the P:O ratio considered (Table 4.1). The latter can be understood by considering that the complete oxidation of the carbon substrate involves TCA cycle function and NADPH generation, because the isocitrate dehydrogenase was assumed to use NADP as cofactor (Table 4.1). The PP pathway had to be used in order to complete the demand for NADPH at P:O ratios of one or more, as for lactate or pyruvate (Table 4.1).

A General Formalism for Metabolic Flux Analysis

In this section, a general formalism to perform MFA developed by Savinell and Palsson (1992a), is presented. It is postulated in algebraic terms of matrices and vectors. The dynamic behavior of the intracellular metabolites is represented by a system of equations:

$$\frac{dX}{dt} = S \cdot v - b \qquad (4.1)$$

X being the vector of n metabolite concentrations, S, the stoichiometry matrix of dimensions $n \times m$, v is the vector of m metabolic fluxes and b the vector of known biosynthetic fluxes.

Under conditions of balanced or steady state growth, the vector of metabolite concentrations may be considered constant, and thus time invariant; so Eq. 4.1 reduces to:

$$S.v = b \qquad (4.2)$$

Usually, the number of metabolic fluxes, m, is much larger than the number of metabolites, n. Among the fluxes, at least a number $(m-n)$ of them must be known (accessible to experimental determination) in order to be able calculate the remaining n fluxes. If the vector of fluxes, v, and the stoichimetric matrix, S, are partitioned between measured and computed fluxes:

$$v = \left| \frac{v_c}{v_e} \right| \qquad S = \left| S_c \, \| \, S_e \right| \qquad (4.3)$$

then Eq. 4.2 becomes:

$$S_c \, v_c + S_e \, v_e = b \qquad (4.4)$$

and v_c can be calculated:

$$v_c = S_c^{-1} \, (b - S_e \, v_e) \qquad (4.5)$$

Therefore, at least $(m-n)$ fluxes have to be measured to be able to estimate the n remaining fluxes. Certainly, the level of accuracy in the estimated n fluxes will depend on the experimental correctness of b and v_e fluxes determination (Savinell and Palsson, 1992a). When the dimension of the vector v_e is lower than $(m-n)$, the system is said to be "underdetermined", and a set of vectors are solutions of the system. In such a case, linear optimization techniques allow to restrict the set of solution vectors (Savinell and Palsson, 1992b).

A Comparison between Different Methods of MFA

Most MFA methods are based on stoichiometric models of metabolism. They were developed in the 90s and have been applied to growth and product formation of microbial and mammalian cells. Although their predictive capabilities are more limited than other modeling techniques, they provide useful information about product yield and flux distribution of large metabolic networks.

MFA Applied to Prokaryotic and Lower Eukaryotic Organisms

The first stoichiometric models of metabolic networks were proposed by Holms and applied to the growth of *E. coli* (Holms, 1986). Later, Stephanopoulos and Vallino (1991) presented the concept of network "rigidity" to explain why the yields calculated from a stoichiometric model of *Corynebacterium* metabolism, to produce the amino acid lysine, were much higher than those attainable under experimental conditions (Stephanopoulos and Vallino, 1991).

Savinell and Palsson (1992a,b) presented a general formalism for the calculation of internal metabolic fluxes based on matrix operations using mass balance equations (see above). These authors named their approach, "metabolic flux balancing" (Varma and Palsson, 1994). Essentially, the method includes linear optimization, both for flux calculation and the sensitivity to experimental error (Savinell and Palsson, 1992a,b). Linear optimization requires an objective function, e.g. the growth rate or the growth yield on a certain substrate, with respect to which the microbial system should find an "optimal" solution (Edwards *et al.*, 2001). To accomplish this task two variables were introduced, namely *shadow prices* and *reduced costs*, that are defined as a function of the sensitivity of an objective function to a constraint (e.g. minimize ATP production or maximize biomass production). Such a method has been applied to bacteria, particularly to *E. coli*, and to hybridoma cell metabolism (Savinell and Palsson, 1992c).

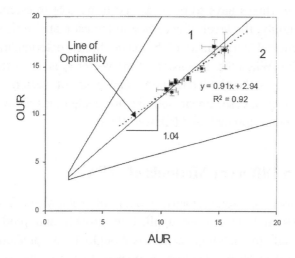

Figure 4.1. *In silico* predictions of growth and substrate consumption rates and comparisons to experimental data. The acetate uptake rate (AUR) versus the oxygen uptake rate (OUR) (both rates in mmol g^{-1} dw h^{-1}) phase plane analysis (phenotype phase plane). All data points lie close to the line of optimality (separating regions 1 and 2) which defines the line of the highest growth yields on both substrates. The errors bars in each data point represent a single standard deviation. Regions 1 and 2 represent non optimal metabolic phenotypes (Reprinted with permission from Nature Publishing Group. Edwards, Ibarra and Palsson, *Nature Biotechnol.* 19, 125-130. ©copyright 2001)

An *in silico* reconstruction of the *E. coli* metabolic network has been performed by Edwards and co-workers (2001) based on a cellular inventory of metabolic gene products. The optimization criterion for maximizing the growth rate was found to be consistent with experimental data from bacteria growing on acetate. Figure 4.1 shows the modeling results as well as the experimental points. The latter were found to correspond to the line of optimality defined by the phenotype phase plane, in this case namely the 3D plot of growth rate as function of both acetate- and oxygen-uptake rate (Edwards *et al.*, 2001). From a physiological point of view, the line of optimality corresponds to a maximal growth yield for both substrates, i.e. acetate and oxygen, under those particular culture conditions.

Sauer and co-workers (1998) applied a similar approach to the production of purine nucleotide, riboflavin or folic acid by *Bacillus subtilis*. The metabolites of interest were all derived from pentose phosphates as intermediary metabolites. Two criteria were used as optimization principles based on: *(i)* a minimal energy expense, and *(ii)* the so called "stoichiometric criteria" that searches to improve the yield of biomass on carbon substrate, minimizing losses as CO_2 and allowing ATP to accumulate (Sauer *et al.*, 1998). The first optimization criterion was obviously dependent on the P:O ratio chosen to perform the calculations and the yield of ATP, Y_{ATP}. The Y_{ATP} parameter adopted in the work of Sauer *et al.* (1998), takes into account the energetic efficiency of the metabolic machinery and encompasses the biomass and product yields on substrate together with the maximal ATP formed per mole of substrate consumed. Y_{ATP} is a useful parameter for distinguishing between an energy-limited biological system and a limitation arisen from the stoichiometry of the pathways involved. As a matter of fact, certain anaplerotic pathways will be activated with final product yields (in moles of product per 6 carbon mole of substrate), varying around 0.6 for guanosine and 0.16 for folic acid, according to the metabolic routes utilized by the carbon substrate (irrespective of whether the uptake mechanisms involve a phosphotransferase system or not).

A matrix metabolic network method was validated through analysis of the growth of *S. cerevisiae* in mixed substrate media, with glucose and ethanol (Vanrolleghem *et al.*, 1996). MFA was performed at the steady state for 98 metabolites, the elemental balance, and 99 reaction rates, including both conversion and transport reactions. According to the glucose fraction in the feed, four modes or regimes of functioning of the metabolic network were observed; the limits between them being predicted by the numerical results. At large glucose fractions, the AcCoA pool was mainly derived from glucose, via pyruvate dehydrogenase, and from ethanol, through the AcCoA synthetase. When the ethanol fraction increases, anaplerotic reactions are fed via a glyoxylate shunt. At even lower glucose fractions, ethanol-derived intermediates enter gluconeogenesis and PEPCK becomes operative. Finally, gluconeogenesis and the pentose phosphate pathway function with ethanol-derived metabolites, above a given ethanol fraction in the feed. Interestingly, the glucose fraction corresponding to the switch between the two metabolic regimes varied if the estimation was made based on a C-balance alone, or if both carbon and cofactor balance were taken into account. The method was validated by a statistical treatment of experimental measurements of fluxes and *in vitro* enzyme activities (Vanrolleghem *et al.*, 1996). The main outcome of the model was the estimation of two important physiological parameters: the P:O ratio and the maintenance energy, i.e. the amount of energy dissipated not associated with growth. Operational P:O ratios ranging from 1.07 to 1.11 mole ATP/O and maintenance between 0.385 and 0.445 mole ATP/C mol biomass, were obtained. The latter value was interpreted by the authors to be independent of C-substrate supported growth.

MFA as Applied to Studying the Performance of Mammalian Cells in Culture

Zupke and Stephanopoulos (1995) have applied MFA to a hybridoma cell line based on material balance of the biochemical network of reactions leading to the synthesis of biomass and antibodies. The method is based on a matrix expression of fluxes at a pseudo-steady state. The calculation was validated by ^{13}C NMR measurement of lactate production. The validation was performed through labeling experiments with Glucose-1-^{13}C and analysis of the label recovery in lactate, focusing on the label distribution among its three carbons. They found reasonable good quantitative agreement between predicted and determined values of intracellular fluxes in cells consuming glucose and glutamine, confirming the assumed biochemistry of the metabolic network.

Vriezen and van Dijken (1998) emphasized the calculations of maximal fluxes through the main catabolic pathways in a myeloma cell line grown in chemostat cultures under several experimental conditions: *(i)* glutamine-limited, which is equivalent to an energy limitation (i.e. mammalian cells in culture drive amino acids into the TCA cycle for complete combustion), *(ii)* glucose-limited, and *(iii)* oxygen-limited. The measured fluxes were confirmed by measurements of enzyme activities involved in the central metabolic pathways of carbon and nitrogen metabolism (Vriezen and van Dijken, 1998).

An alternative mass balance method based on a stoichiometric matrix was applied by Bonarius *et al.* (1996) to the growth and antibody production of hybridoma cells in different culture media. The flux distribution was found using the Euclidean minimum norm as an additional constraint to solve an undetermined system (Bonarius *et al.*, 1996).

Cybernetic models, introduced by Ramkrishna and coworkers (Varmer and Ramkrishna, 1999), are based on the hypothesis that metabolic systems have evolved optimal goal oriented strategies. This modeling approach has been applied to the growth of *S. cerevisiae* (Giuseppin and van Riel, 2000). It uses linear optimization toward several objectives through cost functions, among which the maintenance of homeostasis is of utmost importance. The main difference with flux balancing analysis is that the assumption of steady state or balanced growth condition is eliminated allowing the calculation of time-dependent phenomena. At each integration step the optimal rates have to be calculated using a constrained optimization algorithm (Giuseppin and van Riel,

2000). The behavior of a yeast culture subjected to glucose pulses has been studied and the objective of maximizing glucose uptake was found to dominate over the other objective functions during the transient (Giuseppin and van Riel, 2000).

This survey does not pretend to be exhaustive and some outstanding papers may have not been included, but its purpose is to illustrate how extended are the uses of metabolic stoichiometric matrix methods, along with their potentiality for diagnosis of limitations in an optimization program.

Metabolic Fluxes during Balanced and Steady State Growth

The regulation of the degree of coupling between catabolic and anabolic fluxes, in turn modulates the amount of flux redirection towards microbial products (see Chapter 3). Microbial products may be divided into two groups: *(i)* fermentation products, and *(ii)* proteins and polysaccharides. In the first group, redirection of fluxes towards fermentation products may be generated as a result of alteration in the redox and phosphorylation potentials (Aon *et al.*, 1991; Verdoni *et al.*, 1992; Cortassa and Aon, 1994; Aon and Cortassa, 1997). During transitions from aerobic to oxygen-limited or anaerobic conditions, the correct balance of both, redox (NADH/NAD; NADPH/NADP) and phosphorylation (ATP/ADP) couples (Senior *et al.*, 1972; Kell *et al.*, 1989; Anderson and Dawes, 1990) are germane to the question of the formation of fermentation products.

Fermentation products can be classified on the basis of their relationship with product synthesis and energy generation in the cell (Table 4.2) (Stouthamer and van Verseveld, 1985; Doran, 1995). The first category corresponds to end- or by-products of energy metabolism synthesized in pathways which produce ATP. Those of the second class are partly linked to energy generation, but require additional energy for synthesis. The third class involves the production of antibiotics or vitamins far-removed from central energy metabolism.

Depending of whether the product is linked to energy metabolism or not, equations for q_p as a function of growth rate and other metabolic parameters can be developed.

The rate of product formation in cell culture can be expressed as a function of biomass concentration:

$$r_p = q_p x \qquad (4.6)$$

where r_p is the volumetric rate of product formation, x, biomass concentration, and q_p, the specific rate of product formation. q_p is not necessarily constant during batch culture (Doran, 1995). For products formed in pathways which generate ATP (Table 4.2), the rate of production is related to cellular energy demand. Thus, as growth constitutes the major energy-requiring function of cells, product has to be formed whenever there is growth, so long as its production is coupled to energy metabolism. Under these conditions, kinetic expressions for product formation must account for growth-associated and maintenance-associated production:

$$r_p = Y_{px} r_x + m_p x \qquad (4.7)$$

where r_x is the volumetric rate of biomass formation, Y_{px}, the theoretical or true yield of product from biomass, m_p, the specific rate of product formation due to maintenance, and x, biomass concentration. Equation 4.7 states that the rate of product formation depends partly on rate of growth but also partly on cell concentration. Taking r_x as equal to μx, then:

$$r_p = (Y_{px} \mu + m_p) x \qquad (4.8)$$

Comparison of Eqs. 4.6 and 4.8 shows that, for products coupled to energy metabolism, q_p is equal to a combination of growth-associated and non-growth-associated terms:

$$q_p = Y_{px} \mu + m_p \qquad (4.9)$$

Similar expressions have been derived for the production of extracellular enzymes by *Bacillus*. In this case a linear relationship between r_p and r_x has been demonstrated (Stouthamer and van Verseveld, 1987):

$$r_p = a x_t + b r_x \qquad (4.10)$$

in which a and b are constants. The amount of product formed depends both on the amount of biomass and on its rate of increase. The amount of biomass, x_t, and the growth rate r_x will be dictated by the culture system used (see Chapter 3).

Bioenergetic and Physiological Studies in Batch and Continuous Cultures. Genetic or Epigenetic Redirection of Metabolic Flux

Introduction of Heterologous Metabolic Pathways

The development of arabinose- and xylose-fermenting *Z. mobilis* strains are a conspicuously successful example of metabolic engineering by introduction of heterologous metabolic pathways (Zhang *et al.*, 1995; Deanda *et al.*, 1996). Z. *mobilis* has become an important fuel ethanol-producing microorganism because of its 5 to 10% higher yield and up to five-fold higher specific productivity compared with traditional yeast fermentations. As its substrate range is restricted to glucose, sucrose, and fructose fermentation, the introduction of the ability to ferment pentose sugars (derived from lignocellulose feedstocks) through metabolic engineering, is a biotechnological achievement of great impact.

Additional advantages of Z. *mobilis* as an ethanol-producing microorganism, are its high ethanol yield and tolerance (up to 97% of ethanol theoretical yield and ethanol concentrations of up to 12% w/v from glucose) along with considerable resistance to the inhibitors found in lignocellulose hydrolysates. Its "generally regarded as safe" (GRAS) character (shared with yeast) allows its use as an animal feed.

Rapid and efficient ethanol production by Z. *mobilis* has been attributed to glucose fermentation through the Entner-Doudoroff pathway which produces only one mol of ATP per mole of glucose. Moreover, reduced biomass formation, and a facilitated diffusion sugar transport system coupled with its highly expressed pyruvate decarboxylase and alcohol dehydrogenase genes, have also been implied (Zhang *et al.*, 1995).

The xylose-fermenting strains of Z. *mobilis* were developed by introducing genes that encode the xylose isomerase, xylulokinase, transaldolase, and transketolase activities (Zhang *et al.*, 1995). Two operons comprising the four xylose assimilation and pentose phosphate pathway genes were simultaneously transferred into Z. *mobilis* CP4 on a chimeric shuttle vector. Enzymatic analysis of Z. *mobilis* CP4 grown in a glucose-based medium demonstrated the presence of the four enzymes whereas these were largely undetectable in the control strain that contained the shuttle vector alone.

The expression of these genes allowed for the completion of a functional metabolic pathway that converts xylose to central intermediates of the Entner-Doudoroff pathway, enabling *Zymomonas* to ferment xylose to ethanol (Fig. 4.2). Xylose is presumably converted to xylulose-5-phosphate and then further metabolised to glyceraldehyde-3-phosphate and fructose-6-phosphate, which effectively couples pentose metabolism to the glycolytic Entner-Doudoroff pathway. The recombinant strain obtained was capable of growth on xylose as the sole carbon source and efficient ethanol production (0.44 g per gram of xylose consumed: 86% of theoretical yield). Furthermore, in the presence of a mixture of glucose and xylose, the recombinant strain fermented both sugars to ethanol at 95% of theoretical yield within 30 hours (Zhang *et al.*, 1995).

Table 4.2. Classification of low-molecular-weight fermentation products.

Class of metabolite	Examples
Products directly associated with generation of energy in the cell	Ethanol, acetic acid, gluconic acid, acetone, butanol, lactic acid, other products of anaerobic fermentation
Products indirectly associated with energy generation	Amino acids and their products, citric acid, nucleotides
Products for which there is no clear direct or indirect coupling to energy generation	Penicillin, streptomycin, vitamins

Reproduced from Doran, P.M. *Bioprocess Engineering Principles.* ©copyright 1995 Academic Press, with permission.

A similar strategy was followed by the same team of researchers to expand the substrate fermentation range of *Z. mobilis* to include the pentose sugar, L-arabinose (Deanda *et al.*, 1996). Five genes, encoding L-arabinose isomerase, L-ribulokinase,   L-ribulose-5-phosphate-4-epimerase,   transaldolase   and transketolase, were isolated from *E. coli* and introduced into *Z. mobilis* under the control of constitutive promoters that permitted their expression even in the presence of glucose. The engineered strain grew on and produced ethanol from L-arabinose as a sole carbon source at 98% of the maximum theoretical ethanol yield, based on the amount of sugar consumed.

Since the plasmid containing the heterologous genes was episomal, the recombinant strain with the ability to ferment L-arabinose exhibited plasmid instability. Only 40% of the cells retained the ability to ferment arabinose after 20 generations of growth in complex media without selection pressure at 30°C, and they completely lost the ability to ferment arabinose within 7 generations at 37°C (Deanda *et al.*, 1996).

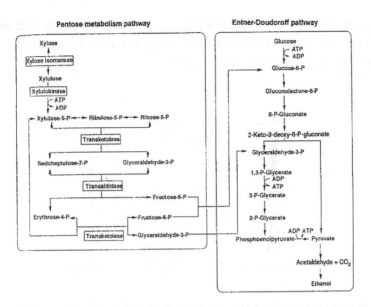

Figure 4.2. Proposed pentose metabolism and Entner-Doudoroff pathways in engineered *Zymomonas mobilis*. The xylose-fermenting strains of *Z. mobilis* were developed by introducing genes that encode the xylose isomerase, xylulokinase, transaldolase, and transketolase activities. The expression of these genes allowed for the completion of a functional metabolic pathway that converts xylose to central intermediates of the Entner-Doudoroff pathway enabling *Zymomonas* to ferment xylose to ethanol (Reprinted with permission from Zhang, Eddy, Deanda, Finkelstein and Picataggio, *Science* 267, 240-243. ©copyright 1995 American Association for the Advancement of Science).

Metabolic Engineering of Lactic Acid Bacteria for Optimising Essential Flavor Compounds Production

The work of Platteeuw *et al.* (1995) shows how, through a combination of molecular biology studies and the establishment of appropriate environmental conditions, the metabolic path leading to diacetyl production from pyruvate, can be optimized in *Lactococcus lactis*. Diacetyl is an essential flavor compound in dairy products such as butter, buttermilk, and cheese, and in many non-dairy products, where a butter-like taste is desired. The pyruvate pathway leading to diacetyl production in *L. lactis* is shown in Fig. 4.3. From two molecules of pyruvate, diacetyl is formed by an oxidative decarboxylation of the intermediate α-acetolactate by the activity of α-acetolactate synthase. α-acetolactate is further decarboxylated to acetoin by the action of α-acetolactate decarboxylase.

Four enzyme activities are known to metabolise pyruvate under different physiological conditions (Fig. 4.3). *(i)* α-acetolactate synthase, which is active at high pyruvate concentrations and low pH; *(ii)* L-lactate dehydrogenase, with maximal activities at high sugar concentrations and high intracellular NADH levels; *(iii)* pyruvate-formate lyase, which is active at a relatively high pH (above 6) and under anaerobic conditions; and *(iv)* pyruvate dehydrogenase, which is active under aerobic conditions and low pH.

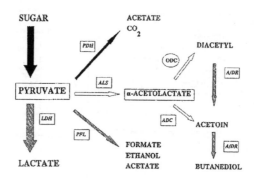

Figure 4.3. Pyruvate pathway in *Lactococcus lactis*. Black and shaded arrows indicate conversions that generate and consume NADH, respectively. Enzymatic and chemical conversions are indicated by boxes and circles, respectively. Abbreviations: ALS, α-acetolactate synthase; LDH, lactate dehydrogenase; PFL, pyruvate formate lyase; PDH, pyruvate dehydrogenase; ADC, α-acetolactate decarboxylase; A/DR, acetoin and diacetyl reductase; ODC, oxidative decarboxylation (Reprinted with permission from Platteeuw, Hugenholtz, Starrenburg, van Alen-Boerrigter and De Vos, *Applied Environmental Microbiology* 61, 3967-3971. ©copyright 1995 American Society for Microbiology).

Taking into account the above metabolic features, strains of *L. lactis* that overproduce α-acetolactate synthase in a wild type MG5276, or in a strain deficient in lactate dehydrogenase, were analyzed for their fermentation pattern of metabolites at different pH and aeration conditions.

Table 4.3. Product formation and lactose consumption by *Lactococcus lactis* MG5267 harboring pNZ2500.

Fermentation condition								Lactose utilized
				Product formed (mM)				
Initial pH	Aeration	Lactate	Formate	Acetate	Ethanol	Acetoin	Butanediol	(mM)
6.8	-	54.6 (100)[b]	N D[c]	ND	N D	N D	N D	18.1
6.0	-	41.2 (99.7)	N D	0.1 (0.2)	N D	N D	N D	9.7
6.8	+	52.0 (70.8)	N D	2.3 (3.1)	N D	9.6 (26.1)	N D	19.2
6.0	+	42.5 (53.4)	N D	2.0 (2.6)	N D	16.1 (42.0)	N D	18.7

Table 4.4. Product formation by lactate dehydrogenase-deficient strain *Lactococcus lactis* NZ2007 harboring pNZ2500.

Fermentation conditions				Product formed (mM)					Lactose utilized
Initial pH	Aeration	Lactate	Formate	Acetate	Ethanol	Acetoin		α-Al[b]	(mM)
6.8	-	3.5 (3.1)[c]	29.1 (25.7)	0.8 (0.7)	24.1 (21.3)	5.5 (9.7)	0.1 (0.2)	22.3 (39.3)	16.6
6	-	2.3 (2.8)	21.9 (26.6)	0.1 (0.1)	17.8 (21.6)	5.5 (13.3)	0.3 (0.7)	14.4 (34.9)	7
6.8	+	1.2 (1.1)	17.6 (15.5)	11.8 (10.4)	10.5 (9.4)	35.0 (61.7)	1.2 (2.1)	ND[d]	20.2
6	+	1.1 (1.1)	ND	5.9 (5.3)	8.7 (7.8)	36.6 (65.7)	0.6 (1.1)	10.5 (18.9)	20.2

Tables 4.3 and 4.4 show the product formation and lactose consumption by both type of recombinant strains of *L. lactis*. The *L. lactis* strain without α-acetolactate synthase (als) overproduction, shows a basal activity of the enzyme. When the plasmid pNZ2500, harboring the lactococcal *als* gene, isolated and cloned under the control of the strong lactose-inducible lacA promoter, was introduced into the wild type strain, an increase of 124-fold in activity over the basal level was observed under inducing conditions in lactose-containing media (Platteeuw *et al.*, 1995).

When cultivated aerobically, *L. lactis* MG5276 harboring pNZ2500 was found to produce acetoin in addition to lactate. At an initial pH of 6.8, 26% of the pyruvate appeared to be converted to acetoin, while at an initial pH of 6.0, almost twice the acetoin production was observed (Table 4.3).

When a lactate dehydrogenase deficient strain (NZ2007) constructed from the lactose-fermenting strain MG5267, was modified with the plasmid pNZ2500 for the *als* overproduction, a range of metabolites were produced under anaerobic conditions (Table 4.4). Approximately half of the pyruvate was converted into formate and ethanol. The major part of the acetoin synthesized was further reduced to butanediol. Under aerobic conditions and an initial pH of 6.8, approximately 64% of the pyruvate was converted mainly into formate (15% of the pyruvate converted), acetate (10%) and acetoin (62%). At a lower initial pH of 6.0, the vast majority (85%) of the pyruvate was converted into products from the α-acetolactate synthase pathway (Table 4.4). These results clearly illustrate that the availability of an electron acceptor, and the intracellular redox status, are main determinants of the fermentation pattern. In agreement with the latter, the transformation of *S. marcescens* with a Vitreoscilla (bacterial) hemoglobin gene shifted the fermentation pathway toward the production of 2,3-butanediol and acetoin (Table 1.1) (Wei *et al.*, 1998). Moreover, the introduction of NADH oxidase into *L. lactis* altered the fermentation pattern from homolactic to mixed acid (Table 1.1) (Lopez de Felipe *et al.*, 1998)

The work of Platteeuw and coworkers (1995) shows that under optimal fermentation conditions, and with appropriately constructed strains of *L. lactis*, up to 85% of the pyruvate (and thereby also the carbon source lactose) was converted via the α-acetolactate synthase pathway to acetoin and butanediol. It also became clear that to produce high yields of diacetyl, the conversion of α-acetolactate into acetoin by the activity of α-acetolactate decarboxylase should be prevented.

Metabolic Control Analysis

This quantitative methodology highlighted the answer to an important question that arises when one deals with networks of biochemical reactions of arbitrary complexity: what steps control the flux through a metabolic pathway? how is the concentration of the intermediaries controlled? It has been mainly applied to systems functioning at steady state.

Figure 4.4 shows a metabolic pathway and its hydraulic analogy. The flux of the fluid (e.g. a carbon source, in grey) flowing through communicating containers is regulated by valves (V1, V2, etc, i.e. enzymes). The level (e.g. amount of intermediary metabolite) attained by the fluid in the containers depends on the aperture of the valves. Intuitively, it is easily seen that valve V3 is the main step controling the flux, since the fluid accumulates to a higher extent in the container controlled by V3. Another important feature is that the level attained by the fluid in each container is different and finite, suggesting that each and every valve controls the flux of the fluid to a certain extent.

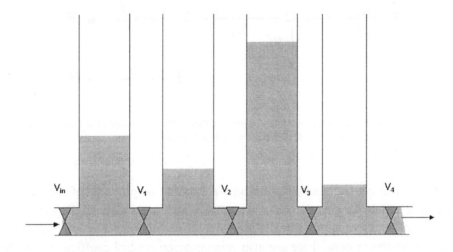

Figure 4.4. A hydraulic analogy of a metabolic pathway.

A hypothetical metabolic pathway and the application of the matrix method of MCA for determining flux and metabolite control coefficients is shown in Fig. 4.5. The metabolic route of Fig. 4.5 represents a network of chemical reactions constituted by branched and linear pathways which share an intermediate B (Appendix A). The purpose of this example is to understand in a pathway of medium complexity, the main basis for the use of MCA. A further experimental example of higher complexity will be described later (see Section: *The TDA approach as applied to the rational design of microorganisms...*).

Considering the hypothetical pathway shown in Fig. 4.5, MCA intends to quantitate two main aspects of its control: that of the flux through that path, and of the intermediary metabolite concentration, through two type of coefficients: the flux control coefficient and the elasticity coefficient. Both type of coefficients represent either global or systemic (flux control coefficient) and local or component (elasticity coefficient) properties of the sequence of reactions represented in Fig. 4.5 (Kacser and Burns, 1973, 1981; Heinrich and Rapoport, 1974; Kacser and Porteus, 1987; see Fell, 1992, and Liao and Delgado, 1993, for reviews).

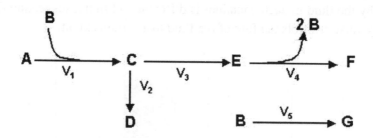

Figure 4.5. A simplified metabolic pathway. The sketched pathway contains the main elements of the glycolytic model shown in Fig. 4.10 (i.e. they are topologically similar).

The control coefficients describe how a variable or property of the system, typically a metabolic flux or the concentration of a metabolite, will respond to variation of a parameter, typically enzyme concentration. For each variable metabolite in the system, there are also concentration control coefficients for the effects of each of the enzymes on that concentration (Fell, 1992).

Figure 4.6 shows different possible relationships between the flux (Ji) of a metabolic route and the activity of a certain enzyme (E_k). In a steady state, the flux control coefficient C_{Ek}^{Ji} is the fractional change in flux for a fractional change in the activity of enzyme, E_k (Kacser and Burns, 1973; Heinrich and Rapoport, 1974):

$$C_{Ek}^{Ji} = \frac{E_k}{J_i} \frac{\partial J_i}{\partial E_k} \tag{4.11}$$

Similarly, a metabolite, Mi, concentration control coefficient, C_{Ek}^{Mi}, can be defined:

$$C_{Ek}^{Mi} = \frac{E_k}{M_i} \frac{\partial M_i}{\partial E_k} \tag{4.12}$$

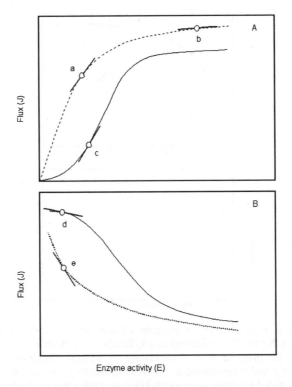

Figure 4.6. Flux (Ji)-enzymatic activity (E_k) relationships, illustrating positive or negative control coefficients. (A) The flux increases with E_k in a hyperbolic (A, a, b) or sigmoidal (A, c) fashion. The flux control coefficient is determined at *a*, *b*, or *c* points, with positive values in all cases. In *a* is slightly lower than 1.0; *b*, close to zero; *c*, higher than 1.0.(B) All flux control coefficients are either slightly (*d*) or strongly (*e*) negative.

Two types of control coefficients have been used, a rate-based (v-type) coefficient and an enzyme concentration-based (e-type) coefficient (Liao and Delgado, 1993). The changes in enzyme activity may arise from changes in the enzyme concentration or modification of their kinetic properties, e.g. k_{cat}.

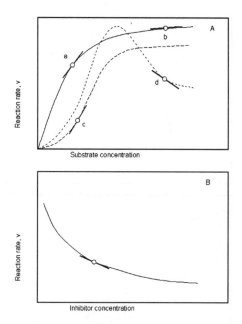

Figure 4.7. Graphical representation of possible relationships between an isolated enzymatic activity (V) and the substrate (S) concentration. (A) Elasticity coefficients are determined from the slope of the curves, the place (*a, b, c, d*) depending on the intracellular concentration of S (see Appendix A). Points *a* and *b* correspond to different saturation states within a Michaelis-Menten relationship while *c* and *d* are from a sigmoidal kinetics. Notice that the higher the saturation, the lower the elasticity coefficient. Point *c* corresponds to an elasticity coefficient higher than 1.0, and point *d* to an enzyme inhibited by substrate with a negative elasticity coefficient. (B) Relationship between enzyme rate and inhibitor concentration; the elasticity coefficient is always negative.

The link between the properties of an enzyme and its potential for flux control is given by the elasticity coefficient, ε. Figure 4.7 displays possible relationships between the rate, v, of an isolated enzymatic reaction and the substrate concentration, S. The elasticity coefficient ε_{ij} for the effect of intermediate metabolite S on the velocity v_i of enzyme Ei is the fractional change in rate of the isolated enzyme, δv_i, for a fractional change, δS, in the amount of substrate S:

$$\varepsilon_S^{vi} = \frac{S}{v_i} \frac{\partial v_i}{\partial S} \tag{4.13}$$

Multiplying by $\dfrac{E_k}{J_i}$ or $\dfrac{S}{v_i}$ makes the coefficients dimensionless and, thus, independent of the units used. All of the terms involved in Eqs. 4.11 and 4.12 are evaluated at steady state.

If an analytic expression for the reaction rate, v, is available, the elasticity coefficients can be obtained by taking the partial derivative of the rate expression with respect to the variable or parameter of interest and then evaluating in the steady state condition (see Appendix A).

Summation and connectivity theorems.

Several theorems constitute the main body of MCA. The summation theorem states that the sum of all flux control coefficients with respect to the activities of each of the enzymatic steps involved in the metabolic pathway being considered, is equal to unity (Kacser and Burns, 1973; Kacser and Porteus, 1987).

$$\sum_k C_{Ek}^{Ji} = 1 \qquad (4.14)$$

This theorem is linked to the concept that the enzymes of the pathway can share the control of the flux. In fact, an important contribution of MCA is to demonstrate that the control of a metabolic pathway can be shared by multiple enzymes, implying that the traditional concept of a single rate-limiting enzyme (a "bottleneck") is inaccurate. This result also illustrates the fact that many rounds of mutation are needed to increase substantially the flux through a pathway (Kacser and Burns, 1981). Moreover, the summation theorem of MCA highlights the fact that the flux control coefficient of an enzyme is not an intrinsic property of that enzyme, but a system property (Fell, 1998). Otherwise, increasing the activity of a rate-controling enzyme also changes its flux control coefficient and so must be the coefficients of enzymes whose activities have not been changed. The latter accounts for the fact that the summation total for the flux control coefficients remains at 1 at all levels of the enzyme whose activity is being increased. This is illustrated in the section below entitled: *"The TDA approach as applied to the rational design of microorganisms…"*

The connectivity theorems relate the elasticities to the control coefficients. For the flux control coefficients, the theorem reads:

$$\sum_i C_{vi}^i \, \varepsilon_k^i = 0 \qquad (4.15)$$

This theorem states that the sum over all products of the flux control coefficients (Eq. 4.11), with respect to the activity of the enzyme catalysing step i, times the elasticity coefficient of this same enzyme (Eq. 4.13), is equal to zero. The connectivity theorem is regarded as the most meaningful of the MCA theorems, for it provides the route to understanding how the kinetics of the enzymes (represented by the elasticities) affect the values of the flux control coefficients (Fell, 1992).

A corresponding set of theorems exist for the concentration control coefficients. The summation theorem states that the sum of all metabolite control coefficients with respect to the activities of each of the enzymatic steps is equal to zero (Kacser and Burns, 1973; Kacser and Porteus, 1987).

$$\sum_i C_{Xi}^M = 0 \qquad (4.16)$$

where M represents any one of the variable metabolites of the pathway, and X_i stands for the activity of enzyme i.

The connectivity theorem becomes slightly more complex, in that it has one form when the metabolite, the concentration of which is the subject of the control coefficient (say A), is different from the one on the elasticities (say B) (Fell, 1992):

$$\sum_{i=1}^n C_{Xi}^A \varepsilon_B^i = 0 \qquad (4.17)$$

but the following form when they are the same:

$$\sum_{i=1}^n C_{Xi}^A \varepsilon_A^i = -1 \qquad (4.18)$$

For unbranched pathways, the summation and connectivity theorems allow the direct calculation of the control coefficients from the elasticities by solving a system of linear algebraic equations (Sauro *et al.*, 1987; Westerhoff and Kell, 1987). In the latter case, the summation theorem and the connectivity theorems for all the metabolites provide exactly the number of simultaneous equations needed for solution of the flux control coefficients of all the enzymes in terms of the elasticities. This is not the case for branched pathways or pathways containing, e.g. substrate cycles (Aon and Cortassa, 1994, 1997; Fell, 1992). For systems involving branches and cycles, additional relationships must be used to solve for the control coefficients from the elasticity coefficients (Sauro *et al.*, 1987; Westerhoff and Kell, 1987; see Liao and Delgado, 1993, for a review). These theorems are the basis of the matrix method, and we will take advantage of them to derive the control of ethanol production by *S. cerevisiae* in continuous cultures (see below) (Sauro *et al.*, 1987; Westerhoff and Kell, 1987; Cortassa and Aon, 1994a,b; 1997; Aon and Cortassa, 1997).

Control and Regulation

It has been argued that a high control coefficient does not necessarily place control at that step, since an enzyme with low flux control coefficient might be significantly controlled through an effector (Fraenkel, 1992). Thus, control is often confused with regulation.

The problem may be addressed through the following question: Do the flux control coefficients indicate which steps "regulate" the flux? (Liao and Delgado, 1993). Conventionally, enzymes sensitive to feed-back inhibition or allosteric effects are said to be the "regulating" enzymes, because without these mechanisms the flux will not be responsive to the change of metabolite pools. In this sense, elasticity coefficients rather than the control coefficients will better reveal the kinetic properties around the point of evaluation. Thus, the flux control coefficients contain no information about the "regulating" enzymes in the above sense (Liao and Delgado, 1993). For enzymes that exhibit high Hill coefficients (a common way of "regulation"), the flux control coefficients tend to zero, while a saturated enzyme (no control in the typical biochemical sense) tends to have a large control coefficient (Kacser and Burns, 1973; Fell, 1992). Therefore, it has been suggested that "control" be used to mean the effect on flux produced by the change of enzyme activity, whereas "regulation" be used to denote the modulation of enzyme activity by effectors (Hofmeyr and Cornish-Bowden, 1991; Fell, 1992; Liao and Delgado, 1993). It has been also emphasized that "control" means that if a change in the level of X occurs, a change in the rate of a process or the amount of, e.g. a metabolite, will follow. This does not necessarily imply that amounts or rates are regulated by X *in vivo*; it will only regulate if a change in X actually occurs in a physiological process (Brown, 1992; Cortassa and Aon, 1994b).

Some mechanisms such as enzyme ambiquity, e.g. hexokinase, an enzyme that can function differently depending on its subcellular location, have been shown to influence the *transient* behavior of glycolytic flux and intermediates. On the other hand, the *steady state* flux of glycolysis was not affected even after large changes in the amount of the mitochondrial-bound form of the kinase (Aon and Cortassa, 1997). After glucose pulses in tumour cells, dynamic ambiquity of HK was able to regulate the transients, although not the steady state control of the glycolytic flux. One important result of the mutual regulation of glycolysis and mitochondrial phosphorylation through HK ambiquity, is that it appears as a biochemical device able to regulate transients more effectively than it can control steady state behavior. Under those conditions, HK ambiquity behaves as a regulatory mechanism exerting feedback on glucose phosphorylation through mitochondrial pools of ATP (Aon and Cortassa, 1997). In connection with these results, it had early been stated by Heinrich *et al.* (1977), that the transient control had to be distinguished clearly from the control parameters for the steady state. Enzymes having a great influence on the transition time (i.e. that for interconversion of substrate to product at that step) may have no relevance for the flux control in the steady state and vice versa (Heinrich *et al.*, 1977).

The Control of Metabolites Concentration

In metabolic networks, some metabolites are not only subjected to mass conversion steps but also play a part in information-carrying networks (Fig. 1.6). Examples of the latter are second messengers and allosteric effectors that being produced at a certain metabolic block, exert their effects in another. For instance, 3PGA that is synthesized in the Calvin cycle in plants, is a positive allosteric effector of ADPglucose pyrophosphorylase (ADPGlcPPase) in the synthetic pathway of starch or glycogen (in plants or cyanobacteria, respectively) (see Chapter 6).

The regulatory role of certain metabolites makes it relevant to know how their intracellular levels are controlled. Elsewhere we attempt to point out some general rules governing the global behavior of metabolic networks through application of MCA. One of the main outputs of MCA is the metabolite concentration control coefficient (see Chapter 7, Table 7.1). An important observation is that metabolite concentrations are controlled by the rate-controling steps of the flux, and negatively controlled by the enzyme consuming the metabolite under analysis. Moreover, whenever a metabolite participates in more than one step or pathway, it is additionally controlled by the enzymes catalysing both or all of those steps.

In Chapter 6 and in the previous section dealing with control and regulation of metabolic networks, we emphasize, with the example of starch synthesis, that an enzyme such as ADPGlcPPase may effectively function both as a mass-energy conversion catalyst, and as an information-transducer device. The latter has important practical applications. When fully activated, an ultrasensitive enzyme such as ADPGlcPPase (Gomez-Casati *et al.*, 1999, 2000) is able to become a rate-controling step (Cortassa *et al.*, 2001; Aon *et al.*, 2001), thus its overexpression is likely to produce increased levels of starch (Stark *et al.*, 1992).

A Numerical Approach for Control Analysis of Metabolic Networks and Nonlinear Dynamics

The description of a particular network of reactions or processes implies the characterization of all possible steady states that may be attained by this particular network (Aon and Cortassa, 1997). Dynamic Bifurcation Theory allows to analyze and characterize the qualitative dynamic behavior of any system of equations in terms of sort and stability of steady states exhibited by that system as a parameter (called bifurcation parameter) is varied (Kubicek and Marek, 1983; Doedle, 1986; Abraham, 1987; Aon and Cortassa, 1997). When plotted, the latter renders a bifurcation diagram (see e.g. Figs. 1.7, 1.8, 5.3, 5.14). The parameter values for which the stability properties change (e.g. from stable to unstable steady states) are called bifurcation points. A visualization of the ensemble of steady states and the localization of the bifurcation points may be obtained with Dynamic Bifurcation Analysis (DBA).

We have applied DBA based on the numerical analysis of ODEs that was performed with AUTO, a software designed for numerical analysis of non linear systems (Doedle, 1986). This numerical procedure allows to compute: *(i)* branches of solutions of stable or unstable steady states, either asymptotic or periodic; *(ii)* evaluation of the eigenvalues and accordingly the stability of such branches; *(iii)* location of bifurcation points, limit points, Hopf bifurcation points and computation of the bifurcating branches (Aon and Cortassa, 1997).

DBA makes possible the analysis of the steady state behavior of a system continuously as a function of a bifurcation parameter. Thus, the flux or concentration control coefficients of MCA can be obtained straightforwardly from the first derivative of bifurcation diagrams plotted in double logarithmic form. Otherwise stated, the slope of the log-log form of the bifurcation diagram allows a graphical and ready estimate of the control exerted on the flux or a metabolite concentration, by a parameter under study (Aon and Cortassa, 1997). All considerations related to the calculation of the kinetic control structure also hold for the thermodynamic analysis of energy transducers based on thermodynamic functions such as dissipation and efficiency (Aon and Cortassa, 1997).

The TDA Approach as Applied to the Rational Design of Microorganisms: Increase of Ethanol Production in Yeast

It has been emphasized (see Chapter 1) that the TDA approach allows us to integrate several disciplines (i.e. microbial physiology and bioenergetics, thermodynamics and enzyme kinetics, biomathematics and biochemistry, genetics and molecular biology) into a coordinated scheme, as represented by the flow diagram of Fig. 1.2. This approach provides the basis for the rational design of microorganisms or cells in a way that has rarely been applied to its full extent.

In the following sections we attempt to further clarify the TDA approach for the particular case of ethanol production in *S. cerevisiae*.

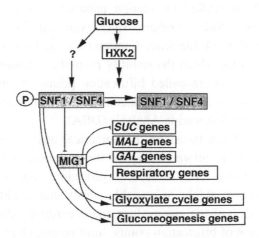

Figure 4.8. Carbon source-dependent gene regulation of glucose repressible genes. In *S. cerevisiae* the *SNF1* (sucrose nonfermenting) gene encodes a serine/threonine protein kinase (72 kDa) (Celenza and Carlson, 1986). Snf1 is physically associated with a 36-kDa polypeptide termed Snf4 which is also required for the expression of many glucose repressible genes and is thought to function as an activator of Snf1. The Snf1/Snf4 protein kinase complex in its active, probably phosphorylated, form is essential for the derepression of possibly all glucose repressible genes (Celenza and Carlson, 1989; Entian and Barnett, 1992; Gancedo, 1992; Ronne, 1995). Under conditions of glucose repression, the Snf1/Snf4 protein kinase is inactive.

Phase I: Physiological, Metabolic and Bioenergetic Studies of Different Strains of S. cerevisiae.

According to the flow diagram of Fig. 1.2, the first step in the TDA approach is to investigate the conditions under which yeast produces the maximal output of ethanol. A comparison between two strains, a wild type and a *mig1* mutant, is shown. A broader screening was performed between several isogenic strains carrying disruption genes involved in the expression of glucose-repressible genes namely *SNF1*, *SNF4*, and *MIG1*, for comparisons with the wild-type strain CEN.PK122. *SNF1* (*CCR1*, *CAT1*) and *SNF4* (*CAT3*) code for proteins that exert a positive regulatory function for the derepression of gluconeogenic, glyoxylate cycle, or alternative sugar-utilizing enzymes as shown in the scheme of Fig. 4.8.

It has been shown that the onset of fermentative metabolism depends upon the glucose repression features of the strain under study (Cortassa and Aon, 1998). Isogenic yeast strains disrupted for *SNF1*, *SNF4*, and *MIG1* genes were grown in aerobic, glucose-limited, chemostat cultures, and their physiological and metabolic behavior analyzed for comparisons with the wild type strain (Cortassa and Aon, 1998; Aon and Cortassa, 1998). Under these conditions, the wild-type strain displayed a critical dilution rate (Dc, beyond which ethanolic fermentation is triggered) of 0.2 h^{-1}, whereas the *mig1* mutant exhibited a Dc of 0.17 h^{-1}. Figure 4.9 describes main physiological variables measured, relevant to the present analysis, e.g. the specific glucose consumption and ethanol production fluxes, and the biomass, as a function of the dilution rate, D. The results obtained with these two strains (i.e. wild-type and *mig1*) are shown since the results showed these to have the highest specific rates of ethanol production. This is in agreement with the aim of the engineering to be performed, i.e. maximal ethanol production. The latter was achieved at high D (=0.3 h^{-1}) (Fig. 4.9), and was higher for the wild-type. In fact, 3.4 and 3 g per liter of ethanol were attained at 0.3 h^{-1} for the wild type and mutant, respectively (see also Table 7.3). Thus, the expected outcome of phase I of TDA has been attained.

Phase II: Metabolic Control Analysis and Metabolic Flux Analysis of the Strain under the Conditions Defined in Phase I

Yields and flux analysis.

MFA may help determine the theoretical as well as the actual yields of ethanol, in order to determine the "ceiling" of the improvement. Otherwise stated, we seek to know the upper limit of increased ethanol production that we may expect by performing the metabolic engineering of the yeast strain chosen, under the conditions already described in chemostat cultures (Fig. 4.9).

For the particular case of ethanol production by yeast, the calculation of the actual as well as potential ethanol yields at high growth rates is quite simple. For the sake of the example, the calculation is given in detail below. Nevertheless, it must be stated that these calculations could be more involved, thereby allowing MFA to show its power. The latter is true, specially when several alternative competing pathways of synthesis of the product desired (e.g. metabolite, macromolecule) are possible. This is the case for carbon flux redirection toward aromatic amino acids synthesis in *E. coli* (Liao *et al.*, 1996; Gosset *et al.*, 1996). Several engineering strategies are possible in this latter case (see Chapter 3).

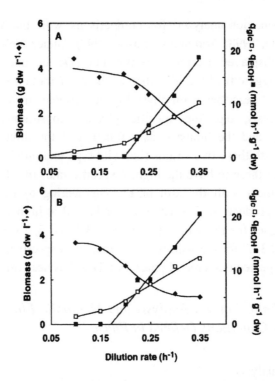

Figure 4.9. Physiological behavior of the wild type strain CEN.PK122 (A) and of the *mig1* (B) mutant in aerobic, glucose-limited chemostat cultures. Biomass yield, glucose consumption and ethanol production rates are plotted as a function of the dilution rate, D, after attaining the steady state judged through the constancy in biomass concentration in the reactor vessel. The values of all physiological variables plotted were obtained from triplicate determinations and the standard deviation was not higher than 5% of the values shown in the figure (Reprinted from *Enzyme and Microbial Technology*, 22, Cortassa and Aon, The onset of fermentative metabolism in continuous cultures depends on the catabolite repression properties of *Saccharomyces cerevisiae*, pp. 705-712. ©copyright 1998, with permission from Elsevier Science).

According to the biomass composition of *S. cerevisiae* under conditions suitable for the purposes of the engineering (Phase I), 7.381 mmole glucose are required for the synthesis of 1 g biomass (see Section: *Stoichiometry of growth....*, Table 4.1). Then, the glucose flux directed to anabolism, i.e. that devoted to replenish key intermediary metabolite precursors of biomass macromolecules, is equal to 7.381 times the growth rate; for a growth rate of 0.3 h^{-1}, this gives 2.214 mmole glucose h^{-1} g^{-1} dw. The rate of glucose consumption was determined to be 7.51 mmol glucose h^{-1} g^{-1} dw (Cortassa and Aon, 1997). The glucose flux not directed to anabolism was then, 7.51—2.214 = 5.296 mmol glucose h^{-1} g^{-1} dw. The yield of ethanol could be therefore calculated from the flux of ethanol production divided by the non-biomass-directed glucose flux (5.793/5.296 = 1.09) with a maximum stoichiometric limit of 2.0 moles ethanol per mole of glucose. Therefore, according to the present analysis, there is still plenty of room for increasing ethanol production, thereby making metabolic engineering worthwhile.

Metabolic control analysis.

Once the growth and metabolic characteristics, and the attainable improvement in product yield have been defined, the task is now to search for the rate-controling steps in ethanol production. The matrix method of MCA (see above and Appendix A) (Sauro *et al.*, 1987; Westerhoff and Kell, 1987) was applied to elucidate the control of glycolysis in *S. cerevisiae* (Cortassa and Aon, 1994a,b, 1997).

The sugar transport and glycolytic model used to interpret the experimental data, are depicted in Fig. 4.10. Important parameters of the model are the gradient of glucose across the plasma membrane, i.e. the difference in concentration of intra- and extra-cellular glucose, and the inhibitory constant, Kin_e, for G6P. Sugar transport in yeast appears to be a significant rate-controling step of the glycolytic flux, and important understanding of the molecular biology of the sugar transporters has been obtained in the past few years (Ko *et al.*, 1993; Reifenberg, 1995; Kruckeberg, 1996; van Dam, 1996). However, it is not clear yet how these different transporters are regulated *in vivo* (Walsh *et al.*, 1996). The equations of the model are described in Cortassa and Aon, 1994, 1997; model parameter optimization and simulation conditions are described in Appendix B.

In order to apply the matrix method of MCA to determine rate-controling steps of ethanol production flux, we need the quantify the intracellular metabolite concentrations required by the model (Fig. 4.10). The latter are introduced into the derivatives of the rate equations, in order to calculate the elasticity coefficients for constructing the corresponding matrix **E**. The inversion of matrix **E** results in the matrix of control coefficients **C**, the first column of which contains the flux control coefficients (see Appendix A).

According to the results obtained by using MCA, in the wild-type at high growth rates, most of the flux control toward ethanol production was exerted by the PFK step (flux control coefficient = 0.93) (Fig. 4.11). A gradual shift of the control from PFK toward HK and the uptake step occurs when the gradient of glucose through the plasma membrane is decreased (Fig. 4.11) (see Cortassa and Aon, 1997; Aon and Cortassa, 1998). Thus, the rate-controling step, the target for genetic engineering, has been identified.

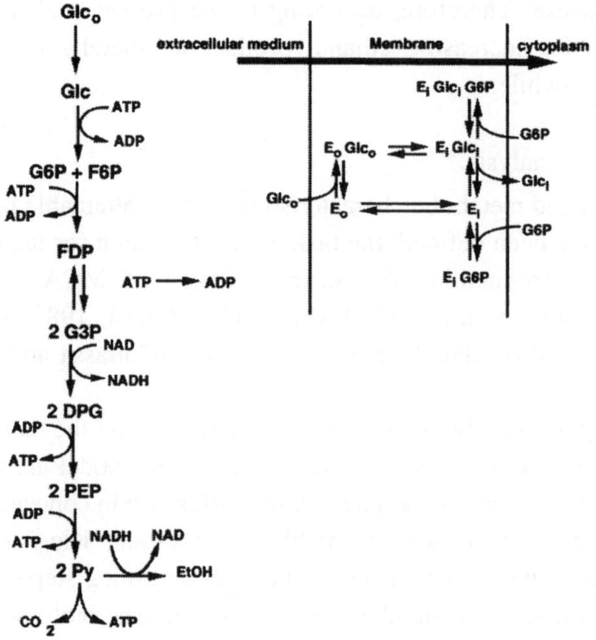

Figure 4.10. Scheme of the glycolytic model and the branch toward the TCA cycle and ethanolic fermentation. Kinetic structure of the model (left). Scheme of the sugar transport step and its inhibition by G6P (right). The transporter molecules bound to G6P, EiG6P, and EiGlciG6P are no longer available for sugar transporting (Reprinted from *Enzyme and Microbial Technology*, 21, Cortassa and Aon, Distributed control of the glycolytic flux in wild-type cells and catabolite repression mutants of *Saccharomyces cerevisiae* growing in carbon-limited chemostat cultures, pp. 596-602. ©copyright 1997, with permission from Elsevier Science).

Phases III and IV: To Obtain a Recombinant Yeast Strain with an Increased Dose of PFK, and to Assay the Engineered Strain in Chemostat Cultures under the Conditions Specified in Phase I

These steps have not yet been attempted for the strain under analysis. Thus we simulated the results with the same model used to interpret the experimental data. We ask whether an effective increase of ethanol production is achieved as can be predicted from MCA. As a first step, we reproduced the experimental steady state values of the glycolytic flux (q_{glc}) and intermediates at $D=0.3$ h^{-1}. Taking this state as the reference one, we increased 10% or 100% PFK (i.e. V_{max}) and obtained 3.8% or 7.2 % increases, respectively, of the flux towards ethanol at the new steady state achieved. For each case above, we tested again whether the control had shifted toward another step. In fact, the control became based on the HK-catalyzed and the uptake steps (0.33 and 0.05 for the 10% increase, or 0.95 and 0.076 for the 100%, respectively).

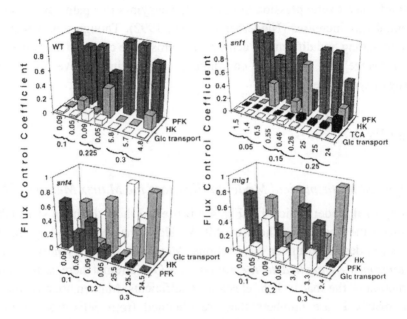

Figure 4.11. Flux control coefficients of the yeast wild type strain CEN.PK122 (WT) and the catabolite-(de)repression mutants *snf1*, *snf4* and *mig1* (B) of yeast run under the conditions described in Fig. 4.8. In all cases, the value of the extent of inhibition constant of the glucose transport by G6P, Kin_e, was fixed at 12 mM while that of the intracellular glucose was allowed to vary, thereby resulting in the glucose gradients indicated (in mM) next to the axis. (Reprinted from *Enzyme and Microbial Technology*, 21, Cortassa and Aon, Distributed control of the glycolytic flux in wild-type cells and catabolite repression mutants of *Saccharomyces cerevisiae* growing in carbon-limited chemostat cultures, 596-602 ©copyright 1997, with permission from Elsevier Science).

Since the increments in ethanol production were relatively modest, we sought further improvement. Thus, we took that steady state characterized by a 7.2% increase of the flux to ethanol as the reference state. As HK is the main rate-controling step in that steady state, we increased the flux through this step by 10% or 100%. With a 10% increase, an additional 4.8% increase in the ethanolic flux could be obtained. Under these conditions, an additional (modest) increase of 20% in the ATPase rate was necessary because of ADP depletion.

Finally we found that a 100% increase in ethanolic fermentation could be theoretically achieved when the fluxes through the glycolytic pathway were increased as follows: through the HK (100%), GAPDH (66%), PGK (73%), ADH (100%), together with the fluxes through the TCA cycle (33%) and the ATPase (100%). It must be stressed that the increases in each of these steps were performed under steady state conditions.

Our results show the feasibility of the fact that, in principle, the simultaneous and coordinated overexpression of most of the enzymes in a pathway can produce substantial flux increases (Niederberger *et al.*, 1992). Though this is technically more demanding and difficult to achieve, it seems to be closer to the way used by cells to change flux levels, i.e. coordinated changes in the level of activity of pathway enzymes (Fell, 1998).

Appendix A

A Simplified Mathematical Model to Illustrate the Matrix Method of MCA

The simplified model depicted in Fig. 4.5 is used to show the main steps taken to implement the matrix method of MCA. How the latter is experimentally achieved, is described under the Sections: *Metabolic Control Analysis* and *The TDA approach as applied to the rational.....* Our attempt here is to show the construction of the matrix **E** of elasticity coefficients. Mathematically, the terms of the matrix **E** are numbers that are obtained from derivation of the rate expressions (e.g. *V*1, *V*2, *V*3) with respect to the substrate (*A* for *V*1) and substrate or effector (*B* for *V*1). Thus, the matrix comprises five reactions (i.e. the columns) and three intermediates B, C, E (i.e. the rows) (Fig. 4.5):

$$
Matrix\ \mathbf{E} = \begin{bmatrix} 1 & 1 & 1 & 1 & 1 \\ ElB1 & ElB2 & ElB3 & ElB4 & ElB5 \\ ElC1 & ElC2 & ElC3 & ElC4 & ElC5 \\ ElE1 & ElE2 & ElE3 & ElE4 & ElE5 \\ 0 & Vr3 & Vr2 & Vr4 & Vr5 \end{bmatrix} \tag{4.19}
$$

The rate equations for each one of the steps in the hypothetic pathway in Fig. 4.5 are as follows:

$$
V_1 = \frac{k1\ A\ B}{A\ B + Ka\ B + Kb\ A + Ka\ Kb} \tag{4.20}
$$

$$
V_2 = \frac{k2\ C}{C + Kc} \tag{4.21}
$$

$$V_3 = \frac{k3\,C}{C + Kcd} \tag{4.22}$$

$$V_4 = \frac{k4\,E}{E + Ke} \tag{4.23}$$

$$V_5 = \frac{k5\,B}{B + Kb5} \tag{4.24}$$

The expression for the derivatives of V_1 with respect to B ($ElB1$), and of the branch at C (V_{r3}), are shown for the sake of the example.

$$ElB1 = \frac{Kb}{B + Kb} \tag{4.25}$$

$$V_{r3} = \frac{k3\,C\,(A\,B + Ka\,B + Kb\,A + Ka\,Kb\,)}{k1\,A\,B\,(Kcd + C\,)} \tag{4.26}$$

The experimentally determined values of metabolite concentrations, are fed into the derivative of the corresponding rate equations (e.g. B for $ElB1$), in order to obtain the elasticity coefficients, which replaced into the explicit expressions of the following **E** matrix:

$$
\begin{bmatrix}
\dfrac{1}{\dfrac{Kb}{B + Kb}} & 1 & 1 & 1 & \dfrac{1}{\dfrac{Kb5}{B + Kb5}} \\[2ex]
0 & \dfrac{Kc}{C + Kc} & \dfrac{Kcd}{C + Kcd} & 0 & 0 \\[2ex]
0 & 0 & 0 & \dfrac{Ke}{E + Ke} & 0 \\[2ex]
0 & \dfrac{k3\,C\,Z}{(C + Kcd\,)k1\,A\,B} & -\dfrac{k2\,C\,Z}{(C + Kc\,)k1\,A\,B} & -\dfrac{k4\,E\,Z}{(E + Ke\,)k1\,A\,B} & -\dfrac{k4\,E\,(Kb5 + B\,)}{(E + Ke\,)k5\,B}
\end{bmatrix}
\tag{4.27}
$$

with $Z = (A\,B + Ka\,B + Kb\,A + Ka\,Kb)$

The matrix **E** is inversed using standard mathematical software (e.g. Maple, Matlab, Mathematica).

Appendix B

Conditions for Parameter Optimization and Simulation of the Mathematical Model of Glycolysis

It is a fair criticism that kinetic parameters of glycolytic enzymes from *in vitro* data could not be representative of their values *in vivo*. In order to address this problem, the steady states achieved by *S. cerevisiae* in chemostat cultures were simulated in the model depicted in Fig. 4.10, as described by ten ordinary differential equations (ODEs) with the aim of optimising the kinetic parameters of the individual enzymatic rate laws. The set of ODEs describes the temporal evolution (dynamics) of ten metabolites of glycolysis as a function of the individual rate laws determined from the enzyme kinetics *in vitro* (Cortassa and Aon, 1994a, 1997). The V_{max} of each rate equation was considered an adjustable parameter. Three criteria were chosen to perform a parameter optimization: *(a)* the system of ODEs should attain a steady state; *(b)* the flux through glycolysis should be equal to the experimentally-measured flux of glucose consumption; and *(c)* the metabolite levels should match the experimentally-determined steady state concentrations. This optimization procedure along with the criteria chosen are very important in order to validate the results of the control of glycolysis *in vivo*. In our hands, it was seen that V_{max} values were the most sensitive to the optimization procedure, rather than K_m, suggesting that the former would be more affected by differences between *in vitro* and *in vivo* data.

Chapter 5

Dynamic Aspects of Bioprocess Behavior

Transient and Oscillatory States of Continuous Culture

Continuous culture, the subject of many laboratory experiments, is not only used as research tool, but is increasingly utilized as a unit process in industry. Examples include the continuous production of beer (Dunbar *et al.*, 1988). In order to ensure the efficient operation of these processes we must understand their unsteady states (dynamic behavior) as well as their steady states. This is necessary so that it may be possible to design effective control systems. Because fluctuations are unavoidable in input variables, the dynamics of chemostat behavior can be unpredictable, but in many cases understanding can help to reduce perturbation to a minimum. It may also be possible to use unsteady states to improve productivity.

Mathematical Model Building

Mathematical model building provides a theoretical framework for understanding mechanisms that underlie the behavior of continuous culture systems, e.g. the importance of hydrodynamic and biological lags in determining the kinetics of responses to altered environmental conditions. Predicted responses made on the basis of idealized models (often derived from chemical reactor studies) can provide new insights into important state variables.

Steps in setting up such a model usually involve:

1. Studies of attainable steady states under specific operating conditions.

2. Determination of whether the behavior of a culture represents a true steady-state, or whether it is characterized by oscillations.

3. Studies of the responses of the system to disturbances.

The simplest case consists of a single-stage, well-stirred, continuous-flow system (CSTR) without recycle or feedback. in practice, continuous cultures are often operated with some type of superimposed control e.g. of turbidity, or of substrate level, by manipulation of dilution-rate. These controls necessarily introduce additional lags, which may impose instabilities not present in the open-loop response. Even though the aim of control strategy is to have more stable operation, simple control systems of the on/off type give rise to overshoot, damped oscillations and off-set problems. More precise control requires more elaborate correction systems (e.g. PID, a proportional integral derivative system). No single model can possibly be adequate to fit the complexity inherent in any microbial system. There is thus always a need to interpret a model system cautiously and verify the major output trends experimentally, and inevitably a model is constructed on the basis of generalized variables. Simplifications arise because the parameters used represent average properties; variations in individual organisms are not usually accounted for. Similarly, the possible presence of sub-populations is neglected. Growth is for the most part treated as a deterministic rather than as a stochastic process. Such assumptions are to some extent valid for large populations of microorganisms, because random deviations from mean values average out these models are sometimes referred to as "unsegregated" and actually represent an abstraction from, and simplification of, real biological behavior. The following example is based on the treatment proposed by Harrison and Topiwala (1974).

Biological systems are thereby regarded as (and described as) homogeneous chemical reactions in terms of vector space-state whose elements (X_i) represent various component concentrations, physical parameters, temperature, pressure, stoichiometric constants etc. The biological rate expression for any material component (y_i) can be expressed as a continuous function:

$$R_j = f(y_1,\ldots,y_k,\ldots T, \text{pressure, mass} - \text{transfer terms, pH}), j = 1, 2,\ldots k \qquad (5.1)$$

R_j is the rate of overall biochemical reaction producing or consuming component y_i per unit volume, per unit time.

A mass balance on any single component y_i the continuous culture yields the equation:

$$\frac{d\,y_j}{dt} = R_j + D(y_j^I - y_j) \qquad (5.2)$$

provided R_j is assumed to be uniformly distributed through the culture, and the growth process has negligible effect on fluid density. Equation 5.2 can be modified to include components present in more than one physical phase to include transport terms. The detail of this kinetic model (i.e. the nature of the R_j terms), determines the theoretical stability of the system. Because microbial growth is so complex, the equations are semi-empirical, and their formulation involves simplification of the real behavior. R_j represents a macroscopic process, and does not describe control systems within an individual cell. Therefore it is difficult to decide the level of complexity of the rate expression necessary for the correct interpretation of the dynamics of the continuous culture system.

In general the model of the continuous culture system as defined by Eqs. 5.1 and 5.2 is represented by a set of first-order, nonlinear differential equations and can be written in matrix notation as:

$$\frac{d\,\underline{x}}{dt} = \underline{f}(\underline{x})$$ (5.3)

where $\underline{x}$ and $\underline{f}$ are n-vectors.

The equilibrium condition of the state-space dynamic model is given by solution of:

$$\underline{f}(\underline{x}) = 0$$ (5.4)

At any point x which satisfies Eq. 5.4 is termed a singular point or steady state. (A solution may not exist, or more than one solution may satisfy the equilibrium conditions: e.g. where substrate inhibition occurs, Andrews, 1968). The stability of the steady state can be found by the Liapounov indirect method (see MacFarlane, 1973). This establishes stability (or instability) of the nonlinear differential equations by examination of stability of the singular point for the locally approximating set of linear equations. To line arise, the model Eq. 5.3 is expanded to a Taylor series about the singular point to obtain:

$$\frac{d\,\underline{x}}{dt'} = \underline{J}\,\underline{x}' + \underline{N}(\underline{x}')$$ (5.5)

where $\underline{x}'$ is the perturbation variable, $\underline{J}$ the Liapounov first approximation matrix and $\underline{N}(\underline{x}')$ a matrix which represents first order terms. The stability of the steady state to small perturbations is determined by the eigenvalue of the matrix $\underline{J}$. A necessary and sufficient condition for a sufficiently small perturbation to die away is that all the eigenvalues of $\underline{J}$ have negative real parts. This represents an overdamped system (Fig. 5.1a).

If an eigenvalue has a positive real part, the steady state will be unstable, and a small perturbation will lead the system away from the steady state (Fig. 5.1b). If $\underline{J}$ has any pure imaginary eigenvalues, the stability will be determined by higher-order terms in $\underline{N}(\underline{x}^1)$. This approach to the steady state does not require exact solution of the nonlinear differential Eq. 5.3. For example in the Monod model of continuous culture:

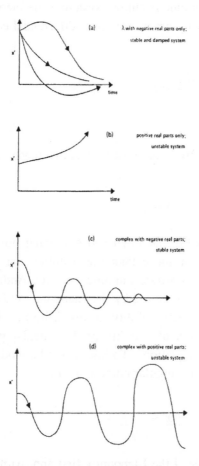

Figure 5.1. Stability of the nonlinear system to a small perturbation of the steady state. Theoretical response as determined by the eigenvalues of the linearized model equations (Reproduced from *Adv. Biochem. Engin.* Transient oscillatory states of continuous cultures. Harrison and Topiwala, vol. 3. 1974 ©copyright Springer-Verlag, with permission).

$$\frac{dx}{dt} = \mu(s)x - Dx$$

(5.6)

$$\frac{ds}{dt} = D(S_R - s) - \frac{\mu(s)x}{Y} \tag{5.7}$$

x is the biomass, s is the substrate concentration in the culture, S_R is substrate concentration in the feed stream (which contains no biomass), μ an arbitrary function of s, is defined as the specific growth-rate, and Y is the constant yield factor.

If x and s represent steady state values, the J matrix for the linearized form of the above differential equations is given by:

$$\underline{J} = \begin{vmatrix} \mu(\bar{s}) - D & \bar{x}\left(\dfrac{d\mu}{ds}\right)_{\bar{S}} \\ -\dfrac{\mu(\bar{s})}{Y} & -\left[\dfrac{\bar{x}}{Y}\left(\dfrac{d\mu}{ds}\right)_{\bar{S}} + D\right] \end{vmatrix} \tag{5.8}$$

The two eigenvalues of the $\underline{J}$ matrix are given by:

$$\lambda_1 = -D \quad and \quad \lambda_2 = \frac{\bar{x}}{Y}\frac{d\mu}{ds_{\bar{s}}} \tag{5.9}$$

Examination of λ_1, and λ_2 yields the condition that the non-trivial steady state ($x>0$, $s<S_R$) will be stable to small perturbations if:

$$\frac{d\mu}{ds_{\bar{s}}} > 0 \tag{5.10}$$

i.e. the steady state will be stable to small perturbation provided that the rate of change of specific growth rate with respect to the substrate is positive, the rate of change being evaluated at the steady state.

For the Monod Model: where μ, the specific growth rate is a simple hyperbolic function.

The non-washout state will always be stable.

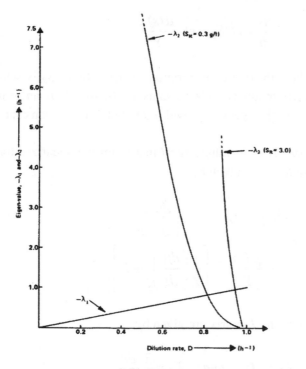

Figure 5.2. Relative importance of the two eigenvalues which determine the speed of response for the Monod chemostat model. Computed numerical values of the eigenvalues plotted as a function of the operating dilution-rate. The values were obtained using the parameter combinations: $S_R = 3$ g l^{-1}; $K_S = 0.012$ g l^{-1}; $\mu_m = 1.0$ h^{-1}. (Reproduced from *Adv. Biochem. Engin.* Transient oscillatory states of continuous cultures. Harrison and Topiwala, vol. 3. 1974 ©copyright Springer-Verlag, with permission).

The eigenvalues of the linearized Monod model can be obtained from Eq. 9 as:

$$\lambda_1 = -D \quad and \quad \lambda_2 = -\frac{(\mu_m - D)[S_R(\mu_m - D) - K_S D]}{\mu_m K_S} \qquad (5.11)$$

where

$$\mu = \frac{\mu_m s}{K_S + s} \qquad (5.12)$$

Since λ_1 and λ_2 consist only of negative real parts for the non-trivial steady state, this model will not give rise to oscillations when subjected to small disturbances. The speed of response of the system as it returns to the steady state will be characterized by two exponentially-decaying modes which will be associated with λ_1 and λ_2 respectively (Fig. 5.2).

Two widely differing modes of response could exist depending on operative dilution rate. λ_1 (representing mixing lag) is much smaller than λ_2 at low dilution rates, but this trend reverses as D approaches μ_m. Important examples of industrial processes which involve inhibition by the substrate can be considered (e.g. waste-water treatment of effluent from a chemical plant which contains toxic xenobiotics).

Stability criteria of Eq. 5.10 are not met when substrate concentration excess of S_1 (Fig. 5.3).

When D does not give washout and feed substrate $S_R > S_1$, this model gives two possible steady states for every value of dilution rate ($D = \mu$), but only that steady state corresponding to the lower substrate concentration and to the left of S, is stable.

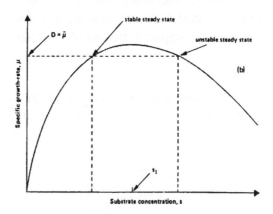

Figure 5.3. Effect on available steady states of substrate/specific growth rate relationship based on Monod-type hyperbolic model, including a substrate-inhibition function. (Reproduced from *Adv. Biochem. Engin.* Transient oscillatory states of continuous cultures. Harrison and Topiwala, vol. 3. 1974 ©copyright Springer-Verlag, with permission).

Transfer-Function Analysis and Transient-Response Techniques

The Liapounov direct method cannot be applied in order to determine responses to finite disturbances from a steady state. A common example is that of a system at start-up from the batch operation to that in continuous mode. A generally applicable method for the determination of which of many possible steady states will be attained is not available. System dynamics can be investigated experimentally by frequency-forcing or pulse testing: e.g. pH, temperature, inlet substrate concentration dilution rate. Lags (mixing and kinetic) limit the usefulness of this approach. The use of sinusoidal variations in input variables is of limited applicability. Pulse-testing is usually more convenient and useful.

Theoretical Transient Response and Approach to Steady State

Continuous culture systems behave in a nonlinear manner, so for large perturbations nonlinear differential equations must be solved. Analytical solutions are rarely feasible and recourse to numerical (computer) solutions is usually necessary (see Chapter 4, the section entitled: *A numerical approach for control analysis and nonlinear dynamics*). Figure 5.3 shows generalized transient responses of the Monod model to stepwise increases in S_R or D. The result is always an overdamped response provided that the system was in steady state prior to disturbances shows a single overshoot or undershoot, depending on whether S_R is increased or decreased. After disturbance, the system returns to a unique steady state. The initial condition of the system does not affect the final steady state. More structured models (i.e. those that account for the 'physiological state' of the culture) give more complex (often oscillatory) behavior.

Substrate-inhibition model.

To study the response of the cell and substrate concentrations after a change in operating conditions the solution of the following differential equations can be considered:

$$\frac{dx}{dt} = \frac{\mu_m\, x\, s}{(K_s + s)\left(1 + \dfrac{s}{K_i}\right)} - D\, x \qquad (5.13)$$

$$\frac{ds}{dt} = D(S_R - s) - \frac{\mu_m\, x\, s}{Y(K_s + s)\left(1 + \dfrac{s}{K_i}\right)} \qquad (5.14)$$

A computer solution of the response of the system starting from two different steady states shows that washout occurs in the case where system is originally in the unstable steady state, but not when system is in the stable steady state prior to the stepwise change.

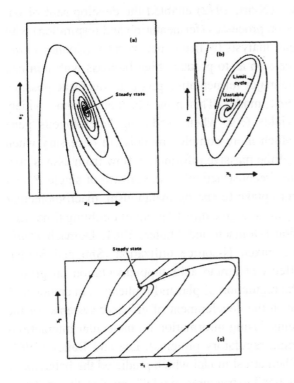

Figure 5.4. (a-c) Examples of generalized phase-planes of state-variables x_1 and x_2 around the steady state. (a) A system displaying damped oscillations as a unique steady state is approached (*Process Analysis and Simulations*, Himmelblau and Bischoff ©copyright 1968 John Wiley & Sons, Inc. Reprinted by permission of Wiley-Liss, Inc., a subsidiary of John Wiley & Sons, Inc.). (b) A system displaying limit-cycle behavior. (c) An overdamped system: no oscillations are obtained as the unique steady state is approached (Reproduced from *Elementary Chemical Reactor Analysis*, Aris ©copyright 1969 Prentice Hall, Inc. Reprinted with permission).

Phase plane analysis.

Many models of continuous culture have been formulated in terms of only two dependent variables (e.g. average cell and substrate concentrations). State variables are plotted against one another through a series of states from initial to final states (Fig. 5.4).

Only trajectories associated with a single steady state or single limit cycle are shown. Cases where multiple states/cycles are also possible occur in more complex examples. Limit cycle behavior indicates that continuous oscillation in state variables occurs, even though temperature, pH, medium flow rate are kept constant. The dynamics of populations in chemostat cultures were simulated in early studies (Tsuchiya, 1983). Soon afterwards, data analysis and computer control of biochemical processes (Nyiri, 1972) enabled the development of on-line balancing of carbon conversion processes (fermentation and respiration) with the growth of yeast, using off-gas analysis by mass spectrometry (Cooney *et al.*, 1977). Optimization of this process so as to produce high biomass with minimal carbon overflow into ethanol, the major fermentation product, was made possible by use of a simple stoichiometric model. This principle then became widely adopted to improve biomass yields (e.g. when producing heterologous gene products in bacteria cultures). Much more complex models are necessary when dealing problems of integration of the massive information flow from assays now available. Early approaches to the "reconstruction" of the growth of single *E. coli* "model bacterium", from nutrient uptake to energy metabolism, macromolecular synthesis and cell cycle traverse, so as to give direct output of doubling times as a function of glucose concentration (Domach and Shuler, 1984; Domach *et al.*, 1984), led on to simulations of mixed N-source utilization (Shu and Shuler, 1989), and prediction of the effects of amino acid supplementation on growth rates (Shu and Shuler, 1991). The regulation of plasmid replication in engineered organisms is an important factor in the development of efficient systems for the production of recombinant proteins. Using information on molecular interactions (e.g. of the affinities DNA-binding regulatory proteins), Lee and Bailey (1984) were able to build a detailed mathematical model which predicted the behavior of the model organism. This "genetically-structured model" enables the mapping from nucleotide sequence, through transcriptional control to the initiation of plasmid replication; it is this that in turn determines plasmid copy number. The principles of Metabolic Control Analysis (see Chapter 4) for the determination of the control properties of individual steps in a pathway and the consequences for metabolic fluxes (Kacser and Burns, 1973; Heinrich and Rappoport, 1974; Heinrich *et al.*, 1977) have recently been extensively applied to problems in biochemical engineering (Varma *et al.*, 1994, Cortassa *et al.*, 1995, Cortassa and Aon, 1994a,b, 1997; Stephanopoulos *et al.*, 1998). Information from *in vivo* NMR analyzes of stable isotope incorporation kinetics into metabolic intermediates currently enables rapid advances based on non-invasively obtained details of metabolic fluxes. Nonlinear dynamics control theory can be incorporated into models in order to predict regions of operation which are characterized by bistability, limit cycle behavior, quasiperiodic or chaotic outputs (Schuster 1988; Lloyd and Lloyd, 1995; Hatzimanikatis and Bailey, 1997; Aon and Cortassa, 1997). Control of complex dynamic behavior offers new approaches to the predictable operation of fermenters driven into their nonlinear operating states (Ott *et al.*, 1990).

Transient Responses of Microbial Cultures to Perturbations of the Steady State

Dilution Rate

Stepwise alteration in the dilution rate, D, is the most commonly employed means of perturbation; this is achieved by raising or lowering the medium feed rate. This imposes on the organism a change in growth rate.

Implicit in the unstructured Monod model (Eqs. 6, 7, 11a) are the principles that: *(i)* growth is regulated *only* by the concentration of growth-limiting substrate, and that *(ii)* microbes possess *all* the constituents necessary for growth at maximum growth rate and can accelerate to maximum growth rate instantaneously when substrate concentration in the culture fluid is raised. These assumptions are not entirely reasonable, because growth processes (metabolism and cell division) are regulated by highly complex control systems. Thus, RNA content increases markedly with growth rate.

Work with Mg^{2+}, K^+ or Pi-limited culture (Tempest *et al.*, 1965) shows that ribosome content regulates protein synthesis rates and hence growth rates. The organism requires a finite adjustment time for the biosynthesis of more RNA before faster growth can occur. In general, only small increases in D can be accommodated almost immediately and without lag. Larger changes in substrate concentration give finite lag times, especially where a substrate is potentially toxic at high concentrations (e.g. methanol). In that case a sequence of reactions occurs whereby a lag leads to a build up of substrate, giving growth inhibition and "washout".

Responses to decreased D will follow the predictions of the unstructured model with regard to cell concentration (because lower growth rate can be achieved instantaneously). However, a finite time will be required for precise readjustment of intracellular conditions (and this depends on turnover times of constituents).

Feed Substrate Concentration

This should have immediate effect on growth rate according to the simple unstructured Monod-type growth model. Effect of sudden increase of substrate (or pulse) is qualitatively similar to increase in D and gives rise to a transient state due to unbalanced growth (Vaseghi *et al.*, 1999). Individual cellular components changing at different rates give rise to change in cellular composition; the RNA content increasing faster than total cell mass. The term "unbalanced growth" often applied in this situation is actually a misnomer: this state is really controlled change between two equilibria states. A small change gives rise to fast accommodation to the new conditions, whereas a large change will give a lag (usually of about 1 to 2 h for a small (1 liter) laboratory fermenter.

Growth with Two Substrates

Cultures containing a mixture of carbon sources have been extensively studied (e.g. glucose and xylose in *E. coli* cultures). In batch culture diauxie occurs with glucose used preferentially before xylose, the substrate that requires an inducible system. In continuous culture, the switch over from glucose to xylose leads to a decreased cell concentration, and to an accumulation of xylose during a transient phase. Then the xylose is metabolised and the cell concentration is re-established. If the nutrient feed is switched back to glucose the result is a smooth transition with no loss of cell density and no accumulation of glucose (i.e. operation of the "constitutive system"). The long recovery times observed (measured in days) indicates that a slow response is required for induction of xylose-metabolising enzymes (which have been previously subjected to glucose repression).

We can conclude that modeling transient responses in chemostats with two substrates must take account of cell physiology, Relaxation times may be expected to be much longer than those observed for single-substrate systems.

Temperature

Change in temperature produces altered growth rates, yield coefficients and affinity for substrate. The content of lipid, carbohydrate and RNA in cells is temperature sensitive. Within limits, responses of steady state growth rates to temperature follows simple Arrhenius relationship. So it would be expected that a sudden increase of temperature applied to a growing culture would give an immediate increase in the maximum growth rate, but in practice a lag is observed before the growth rate is accelerated to the new value. This is because intracellular controls have to be adjusted, new cellular machinery must by synthesised; levels of RNA, regulatory molecules, enzymes, and membrane structure and function all have to be modulated. Decrease in temperature in general produces a smoother adjustment process and consequently a smaller lag. Here again, transient responses indicate complex intracellular regulation.

Dissolved Oxygen

The oxidases (e.g. cytochrome *c* oxidase in yeast, or several different bacterial oxidases) have high affinities for O_2 in organisms that have been grown aerobically. Apparent Km values for O_2 of about 0.1 µM are typical. Thus O_2 the terminal electron acceptor, saturates the oxidase above a low threshold, and over a wide range of concentrations and *does not limit respiration*. So over this range (e.g. 5-280 µM for bakers yeast), the culture is usually completely insensitive to the concentration of O_2. At very high (especially hyperbaric) O_2, almost all microbial species are inhibited, due to formation of reactive oxygen species. At very low dissolved O_2, dynamic responses can be very complex, especially in facultatively anaerobic species; e.g. *Klebsiella aerogenes* (Harrison and Pirt, 1967) or *Saccharomyces cerevisiae* (Lloyd, 1974). Changes occur on differing time scales. Rapid responses (on time scales of minutes) represent metabolic feedback in allosteric enzymes. These are followed by slower changes (measured in hours) that require new enzyme synthesis.

The Meaning of Steady State Performance in Chemostat Culture

So we may ask "*What is meant by the steady-state?*" How long after a perturbation to a continuous culture must we allow to elapse before we can assume that the culture is again at steady state? There is no absolute answer to this question; a true steady state probably never exists in a biological system (otherwise there could be no process of evolution). In most studies the term "steady state" culture is an operational one, and refers to a population of cells kept constant over several generations. There may be *multiple possibilities for steady states*. This is especially so in mixed cultures of microorganisms. For instance, this is a very important principle in waste-water treatment. The actual steady state attained depends on past history of the culture and on the previous perturbations it has experienced. Transitory responses provide the best means of defining regulatory mechanisms involved in cell metabolism e.g. the Pasteur Effect in yeast, mammalian cells, bacteria. When O_2 is decreased, glucose-6-phosphate decreases and then recovers, whilst fructose 1, 6-diphosphate and triose phosphate increase, before decreasing again. This "cross-over point" between those metabolites which show transient decreased concentrations and those that increase, indicates an important site of glycolytic control by phosphofructokinase (Ghosh and Chance, 1964).

Oscillatory Phenomena in Continuous Cultures

1. Oscillations as a Consequence of Equipment Artifacts

Poor feedback control of parameters: (e.g. pH, stirring speed, foam or volume control) can lead to instabilities. Even small oscillations in these environmental factors are amplified to give large fluctuations in other culture parameters. Thus in a continuous culture of *Pseudomonas extorquens* (e.g. at pH 7.0, pH fluctuations of <0.1 unit) the usual feedback system used for pH control (linked to the output of a pH electrode) has been known to give fluctuation of ± 6% total respiration; this in turn can give a fluctuation in dissolved O_2 of as much as ± 25% of air saturation (Harrison and Loveless, 1971). Sensitivity of the cells to small changes in pH varies across the pH range; sometimes when pH is altered even by small amounts, acid fermentation products (acetic and formic acids) replace non-ionisable products (ethanol and acetoin) and the pH of the culture then falls precipitously.

Temperature: especially at upper or lower end of growth range, small changes in temperature can give large changes in physiological activities and growth rates.

Stirring rate fluctuations: affect mixing and especially gas transfer functions (e.g. K_{La}).

Foaming: also affects K_{La}; antifoam agents (surface active agents) affect membrane functions (e.g. cell respiration).

Discontinuous substrate feed: Pulse-feeding of substrate can give oscillations in respiration rate, dissolved O_2 and pH. Also the yield coefficient may vary with pulse frequency even though the average D is kept constant. This observation indicates poor regulation of energy metabolism in these organisms—in the sense that excess substrate is wastefully oxidized. Some feed pumps do not provide steady continuous flow; a peristaltic pump is not as good as a syringe pump activated by stepper motor in this respect. In large-scale processes, components of a single feed stream may not be completely mixed.

2. Oscillations Derived from Feedback Between Cells and Environmental Parameters

Here, feedback between cells and environment, has an effect on the environment, which in turn feeds back on the cells. Thus when the pH of a weakly-buffered culture is not controlled, or if the culture growth produces acid, the pH falls, and the metabolism slows or stops. Then the pH increases again due to influx of fresh medium, metabolic activity increases again, giving damped or continuous oscillations depending on lags in the system. This sequence of events thereby leads to oscillations in respiration and dissolved O_2. Thus it was shown that in *K. aerogenes* the oscillatory state requires three conditions.

(i) At low O_2, respiration rate increases (Degn and Harrison, 1969); whereas

at higher levels of dissolved O_2 (<20 μM O_2), respiration rate is independent of O_2.

(ii) Some metabolic intermediate, a substrate for respiration, builds up and then becomes depleted.

(iii) Oxygen transfer from gas-liquid phase is limited by a low value for K_{La}. Increased stirring rate (i.e. the lowering of the diffusional barrier to O_2) in this case prevents the oscillations.

Growth of *E. coli* in a chemostat can give oscillating pyruvic acid production, and this was observed as spikes which occurred at hourly intervals. Pyruvate initially was produced at a high rate, and then rapidly oxidized. Production repressed at high pyruvate concentration, but derepressed at low pyruvate concentration. Because of time lags in the system, feedback control continuously produced overshoots and this led to undamped oscillations in the concentration of pyruvic acid in the culture.

3. Oscillations Derived from Intracellular Feedback Regulation

Rhythmic phenomena are ubiquitous in biological systems due to physiological control processes. These oscillatory states, may represent "sloppy" regulation, but more importantly serve a variety of functions including timekeeping or signalling (cell-cell and intracellular, Goldbeter, 1996). Where advantageous they are highly conserved during biological evolution as biological rhythms and clocks (Edmunds,1988; Lloyd, 1992).

Glycolytic oscillations.

These are the most frequently studied example of oscillatory states. They were first reported by Ghosh and Chance (1964) who observed NADH oscillations in yeast and measured the fluctuating pool sizes of glycolytic intermediates. Mechanisms were predicted (Higgins, 1967; Sel'kov, 1968) and experimentally validated (Chance *et al.*, 1967). The first demonstration of a sustained glycolytic oscillation in a suspension of intact yeast was that by Von Klitzing and Betz (1970).

Respiratory oscillations.

K. aerogenes grown in continuous cultures sometimes shows NADH and dissolved O_2 oscillations at high frequency, and low amplitude, i.e. (< 1% total respiration rate) (Harrison *et al.*, 1969; Harrison, 1970). These oscillations (Fig. 5.5) have been shown to be insensitive to changes in pH, temperature, O_2 tension, or medium supply rates, and they occurred at high O_2 (i.e. well above "critical" dissolved O_2).

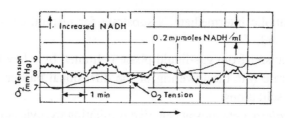

Figure 5.5. Oscillations of NADH fluorescence and dissolved oxygen tension obtained in a glucose-limited chemostat culture of *K. aerogenes* at $D = 0.2$ h^{-1} and pH = 6. 0. (Reproduced from Harrison, 1970 Undamped oscillations of pyridine nucleotide and oxygen tension in chemostat cultures of *Klebsiella aerogenes. J. Cell Biol*, 45, 514-521, with permission).

They were not due to cell-environment interactions, but arose from intracellular feedback loops initiated by an anaerobic shock. Damping was decreased by *repeated* anaerobic shock, although sometimes these oscillations arose spontaneously. It was concluded that there must be some cell-cell communication in order to produce and maintain population synchrony although the putative synchronizing substance has never been identified.

The induction and elimination of oscillation in continuous cultures of *S. cerevisiae* (Parulekar *et al.*, 1986) by varying dilution rate, agitation speed, and dissolved O$_2$, did not provide a mechanistic explanation. A robust autonomous respiratory oscillation ($\tau = 30$-120 min) also occurs in certain strains of acid-tolerant *S. cerevisiae* grown under continuous aerobic culture conditions (Satroutdinov *et al.*, 1992). The oscillation occurs independently of glycolysis, the cell cycle (Keulers *et al.*, 1996b) and no difference in oscillations was observed when cultivation was carried out in light or darkness (Murray *et al.*, 1998). The oscillation is dependent on pH (Satroutdinov *et al.*, 1992), aeration (Keulers *et al.*, 1996a), and carbon dioxide (Keulers *et al.*, 1996a). Oscillation occurs when glucose, ethanol or acetaldehyde is used as a carbon source. For the oscillation to occur ethanol has to be present. It has been suggested that a stage of ethanol metabolism may be a locus for both population synchrony and intracellular regulation giving oscillatory dynamics (Keulers *et al.*, 1996b).

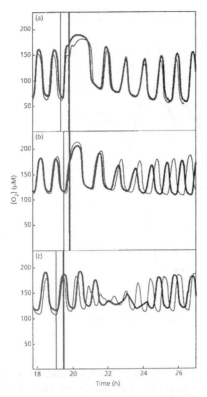

Figure 5.6. Perturbation of dissolved oxygen oscillation during continuous aerobic yeast culture, with GSH (50 μM) (a), GSSG (25 μM) (b), and NF (5-nitro-2 furaldehyde, an inhibitor of GSH reductase) (50 μM) (c). The thick lines represent injection at high dissolved oxygen, the fine lines represent injection at low dissolved oxygen and the vertical bars represent the time of addition (Reproduced from Murray, Engelen, Lloyd and Kuriyama, 1999, Involvement of glutathione in the regulation of respiratory oscillations during a continuous culture of *Saccharomyces cerevisiae*. *Microbiology* 145, 2739-2745 by permission of The Society for General Microbiology).

With ethanol, continuous monitoring indicates an oscillation of NAD(P)H with a complex waveform, and with the predominant period of 45 min identical with that in dissolved O_2 (Murray *et al.*, 1998). Intracellular GSH also oscillates with the same period (Fig. 5.6).

These respiratory oscillations are extremely sensitive to perturbation by pulse addition of Na nitroprusside (Fig. 5.7) and this effect appears to be specifically mediated by nitrosonium ions (NO^+), as NO (gas) or NO-donors are not effective (Murray *et al.*, 1998). Preferred target sites are probably either thiols or protein metal centers.

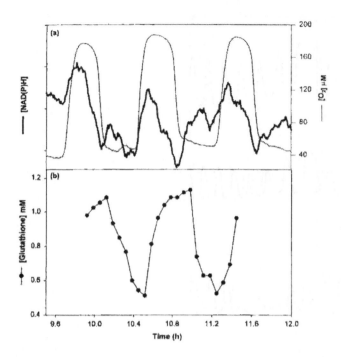

Figure 5.7. Respiratory oscillations of the changing intracellular redox states of yeast. (a) Dissolved O_2 and NAD(P)H fluorescence (365→450 nm) measured continuously on-line. (b) Total intracellular glutathione (Reprinted from *FEBS Lett.* 431, Murray, Engelen, Keulers, Kuriyama and Lloyd. NO^+, but not NO, inhibits respiratory oscillations in ethanol-grown chemostat cultures of *Saccharomyces cerevisiae*, 297-299. ©copyright 1998, with permission from Elsevier Science).

Reduced glutathione (GSH) itself also produces a marked effect on the respiratory oscillations (Murray *et al.*, 1999) leading to an interruption of oscillatory behavior due to respiratory inhibition (Fig. 5.8). No evidence has been found for cell division cycle synchrony in these cultures; the mean division time under the conditions employed is about 12 h. Our interpretation of these data suggests the operation of a respiratory switch during cycling between high and low respiratory activity.

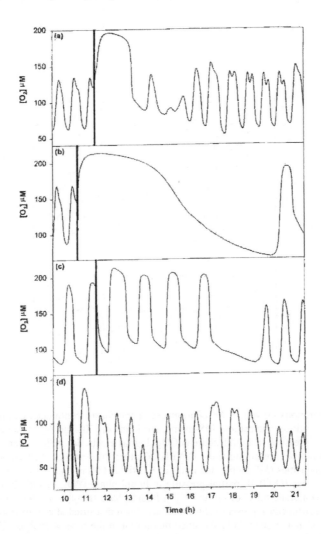

Figure 5.8. Perturbation of respiratory oscillations in a yeast culture. Continuous traces of dissolved O_2 are shown; vertical bars indicate time of addition of (a) 100 μM glutathione, (b) 5 μM Na nitroprusside, (c) 8 μM $NaNO_2$, (d) 10 μM *S*-nitrosoglutathione (Reprinted from *FEBS Lett.* 431, Murray, Engelen, Keulers, Kuriyama and Lloyd. NO^+, but not NO, inhibits respiratory oscillations in ethanol-grown chemostat cultures of *Saccharomyces cerevisiae*, 297-299. ©copyright 1998, with permission from Elsevier Science).

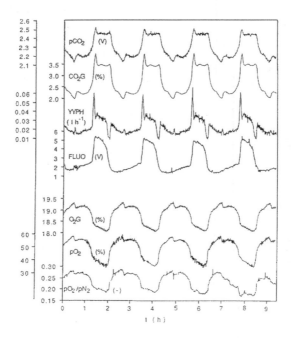

Figure 5.9. Synchronous growth of *Saccharomyces cerevisiae* in continuous culture at a dilution rate of 0.182 h^{-1}: ratio of intensities (MS-membrane inlet) of CO_2 and N_2 in liquid phase (pCO$_2$ /pN2)' Col partial pressure in liquid phase (pCO$_2$, Ingold electrode). CO_2 content in exhaust gas (Co$_2$G). Flow rate of pH controling agent (YYPH). Culture fluorescence (FLUO, arbitrary voltage units). Ratio of intensities (MS-membrane inlet) of ETOH and N_2 in liquid phase (pETOH/pN2)' O_2 content in exhaust gas (O$_2$G)' O_2 partial pressure in liquid phase (PO$_2$, Ingold electrode). Ratio of intensities (MS-membrane inlet) of O_2 and N_2 in liquid phase (PO$_2$/pN$_2$). The gaps in the signals from the mass spectrometer represent values that have been discarded after unsuccessful I test. The ripple in the O$_2$G-signal is due to a badly tuned thermostat in the oxygen analyzer (Reprinted from *J. Biotechnol.* 9, Strassle, Sonnleitner and Fiechter, A predictive model for the spontaneous synchronization of *Saccharomyces cerevisiae* grown in continuous culture. II Experimental verification, 191-208 ©copyright 1989, with permission from Elsevier Science)

Further work is necessary to show whether a control mechanism of the type proposed for *E. coli* (Demple, 1998) may also function here. This continuously oscillating yeast system provides a convenient model, which may be analogous to longer-period redox cycling systems; temperature compensation of period (Murray and Lloyd, unpublished) indicates a timekeeping function (ultradian clock).

Growth rate oscillations.

Dean and Moss (1970) showed that barbiturate inhibition of *K. aerogenes* induced oscillations in growth rates in a turbidostat culture. Uptake of the drug was dependent on growth rate and this phenomenon gave a highly damped oscillation with a period of about 2 generation times.

4. Oscillations Derived from Interactions between Different Species in Continuous Culture

Protozoa and bacteria grown together in continuous cultures often show the oscillatory predator/prey population relationships under constant environmental conditions e.g. with glucose as the limiting nutrient (Curds and Cockburn, 1971). In this system variations in the temperature conditions or residence times gave a reasonable fit to a model, except that damping eventually occurred (perhaps due to wall-growth which would be expected to stabilize the system).

This approach to growth dynamics of mixed cultures has proved useful for the modeling of microbial interactions in natural environments e.g. in sludge-treatment plant, soil, sediments and in medical microbiology. Developments include the modeling of the spread of antibiotic resistance through and between populations and an understanding of epidemiology (e.g. the spread of human disease; AIDS, malaria and measles).

5. Oscillations Due to Synchronous Growth and Division

To study synchronized population growth and division in continuous culture, periodic starvation and reseeding regimes have been most valuable, e.g. in *Candida utilis* and in *E. coli* (Dawson, 1985). "Spontaneous" synchrony described by Kuenzi and Fiechter (1969), and Von Meyenburg (1973), has been analyzed in continuous cultures of *S. cerevisiae* by a number of groups including those in Zurich and Milan (Strässle *et al.*, 1988, 1989; Porro *et al.*, 1988). In the former, investigations of a number of variables were simultaneously monitored (Fig. 5.9) and it was shown that oscillatory behavior of O_2 consumption, CO_2 production, ethanol production, NADH fluorescence intensity and biomass results from spontaneously generated synchronisation of cell division cycles in the entire population of organisms.

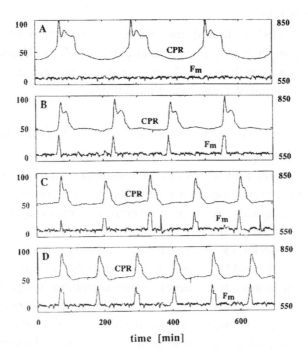

Figure 5.10. The subplots A-D show a spontaneously synchronous (A) and three forced synchronous cultures (B-D) of *S. cerevisiae* at the constant dilution rate $D = 0.13$ h^{-1} at pH 4.0. The forcing period varies from 162 (subplot B), 132 (C) down to 112 (D) min. The spontaneous oscillation period is 215 min. F, is the medium flux (measurement value), CPR is the carbon dioxide production rate. The amount of the pulses in the forcing function was 60 mg l^{-1} glucose. (Reprinted from . *J. Biotechnol.* 24, Münch, Sonnleitner, and Fiechter. New insights into the synchronization mechanism with forced synchronous cultures of *Saccharomyces cerevisiae*, 299-314. ©copyright 1992, with permission from Elsevier Science).

More recent analyzes have shown that in "forced" synchrony cultures (Fig. 5.10) the period and degree of synchrony of spontaneous oscillations can be manipulated by repetitive small pulses of carbon and energy source (Münch *et al.*, 1992).

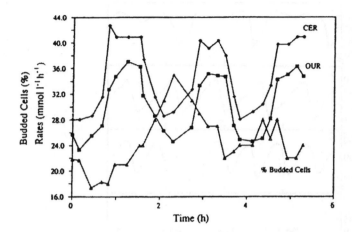

Figure 5.11. Cyclic carbon dioxide production, oxygen uptake and % budded cells for synchrony oscillations at 0.14 h^{-1} in the calorimeter. The budded cell curve is out of phase with the CER and OUR. (Reprinted from *J Biotechnol* 29 Auberson, Kanbier, and von Stockar. Monitoring synchronized yeast cultures by calorimetry. 205-215 ©copyright 1993, with permission from Elsevier Science).

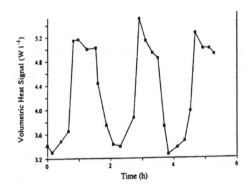

Figure 5.12. Cyclic volumetric heat production rates for synchrony oscillations at 0.14 h^{-1} in the calorimeter. With this smaller time scale, it can be seen that the period for one cycle is 1.8 h. (Reprinted from *J Biotechnol* 29 Auberson, Kanbier, and von Stockar. Monitoring synchronized yeast cultures by calorimetry, 205-215 ©copyright 1993, with permission from Elsevier Science).

The periodic nature of respiro-fermentative metabolism have been further analyzed by calorimetry (Figs. 5.11, 5.12) (Auberson *et al.*, 1993), and by cell component analysis (Duboc *et al.*, 1996).

Systematic analysis of oscillations as a function of two important controling parameters, dilution rate and dissolved O_2 (Figs. 5.13, 5.14) have been made by Porro *et al.* (1988).

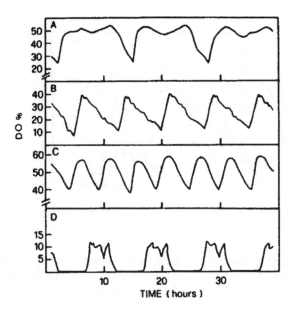

Figure 5.13. Oscillation of dissolved oxygen (DO%) in continuous cultures of *S. cerevisiae* at different dilution rates: (A) D = 0.07 h^{-1}, agitation speed 200 rpm; (B) D = 0.106 h^{-1}, 200 rpm; (C) D = 0.15 h^{-1}, 300 rpm; and (D) D= 0.166 h^{-1}, 200 rpm (Oscillations in continuous cultures of budding yeast: A segregated parameter analysis. Porro, Martegani, Ranzi and Alberghina, *Biotech. Bioengin.* 32, 411-417. ©copyright 1988 John Wiley & Sons, Inc. Reprinted by permission of Wiley-Liss, Inc., a subsidiary of John Wiley & Sons, Inc.).

In a new model for spontaneous oscillations in yeast cultures alternate growth on limiting glucose and limiting glucose plus ethanol is invoked (Martegani *et al.*, 1990). More complex behavior in continuous cultures controlled by ac impedance feedback includes fluctuations in growth rate (Davey *et al.*, 1996). A variety of time series analyzes (Fourier transformations, determination of Hurst, Lyapunov and embedding dimensions as well as nonlinear forecasting techniques) were employed to demonstrate the presence of a chaotic attractor in the dynamics.

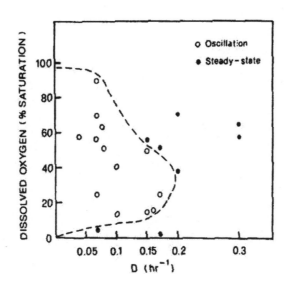

Figure 5.14. Existence of oscillations as a function of D and DO% in glucose-limited continuous cultures of *S. cerevisiae* (Oscillations in continuous cultures of budding yeast: A segregated parameter analysis. Porro, Martegani, Ranzi and Alberghina, *Biotech. Bioengin.* 32, 411-417. ©copyright 1988 John Wiley & Sons, Inc. Reprinted by permission of Wiley-Liss, Inc., a subsidiary of John Wiley & Sons, Inc.).

Chapter 6

Bioprocess Development with Plant Cells

Microorganisms were the first biological systems employed in MCE. The relatively ease of manipulation of many bacteria and of a few fungi (typically *S. cerevisiae*), including short generation times, as well as the information available on their genetics and physiology, have contributed to the preferred choice of microorganisms for MCE. This is well mirrored in the previous chapters of this book, where mathematical approaches, experimental methods, and designs, are exemplified and derived from studies mainly performed with microorganisms.

Unlike microorganisms, plant cells are more complex in structure, physiology and metabolism. In addition, for many years, biochemical studies on metabolism were concentrated on bacteria and animal cells; with the idea that what happens in a plant cell should be very similar to that occurring in the former. This is clearly observed in most textbooks on biochemistry, where metabolism of animals and microorganisms are exhaustively considered; whereas distinctive fundamental aspects of plant biochemistry receive no mention. Despite this, several features of plants determine that they are currently considered as the most promising organisms to be used in MCE. In fact, the production of transgenic plants with increased resistance to herbicide, insect, or viruses shows that, in this area, the development of plant biotechnology is more advanced than in that achieved for animals. For these reasons, it is clear that the complete characterization of plant metabolism and physiology, followed by accurate studies of matter and energy balance and cellular process quantitation are highly relevant issues at the present time.

In this chapter, we consider the use of plants as biological systems for MCE. We focus on the current and potential resources in the field, and discuss the distinctive problems arising from the complexity of plant cells for the rational study of metabolism and metabolic engineering. The analysis involves both the plant cell and the whole organism, and presents a challenge for the rational study of metabolism at higher levels of complexity.

MCE in Plants: Realities and Potentialities

Many distinctive characteristics make plants a unique target system for MCE. Mostly based on autotrophy towards the assimilation of inorganic CO_2, photosynthetic organisms are the most important renewable resource commodity on the earth (Owen and Pen, 1996). Plants have been historically used to satisfy demands for food, as well as to obtain raw materials for non-food industries (including chemicals and pharmaceuticals) (Somerville and Bonetta, 2001). The emerging techniques of genetic transformation have opened relevant possibilities for their application in plants, not only in order to improve the productivity of traditional crops, but also to obtain organisms synthesizing novel products. Thus, a vision of plants as natural bioreactors, having the advantage of producing high quality biomass at relatively low cost, is increasingly considered. Many goals have already been reached in the area, and the future in this discipline is highly promising (Collins and Shepherd, 1996; Owen and Pen, 1996; Birch, 1997; Lundblad and Kingdon, 1999; Ohlrogge, 1999; Willmitzer, 1999; Della Penna, 2001; Somerville and Bonetta, 2001).

Approaches for metabolic engineering as related to plants are many-faceted. A schematic division of purposes and objectives of plant transformation by genetic engineering is illustrated in Table 6.1. Manipulation of genetic information in plant cells is a very important tool that can primarily be utilized for basic or applied research (Birch, 1997; Della Penna, 2001; Somerville and Bonetta, 2001). Obviously, applied strategies are dependent on the former; mostly because the use of this methodology in basic research results in the better understanding of physiological processes; this fundamental work feeds potential new applications. In fact, a good understanding of plant metabolism at a basic level, including the knowledge of metabolic fluxes and their regulation, is an essential prerequisite for the rationale transformation of a plant in a MCE framework.

Plant Transformation for Studies on Metabolism and Physiology

The possibility of modifying the expression (including both over- or under-expression) of specific genes in plants contributes to the study of their metabolism and physiology by different (new and powerful) ways than those allowed by other biochemical methods. Table 6.1 shows examples of key plant processes that are being studied by using this tool. These studies are providing key information for potential application to the improvement of plant products. Firstly, the analysis on the role of certain enzymes in the allocation and partitioning of photoassimilates is relevant for the understanding of factors determining the harvestable yield of a crop; this is critical for rational manipulation and improvement (Stitt, 1994; Herbers and Sonnewald, 1996; Somerville and Bonetta, 2001). Secondly, the characterization of the involvement of specific enzymes and hormones in plant development is important for the determination of factors altering the postharvest maintenance of fruits and flowers (Theologis, 1994; Mol *et al.*, 1995). Finally, the study of plant-microbe interactions and their dependence on cellular signals has allowed us to gain key information about the understanding of natural plant resistance to pathogens (Staskawicz *et al.*, 1995).

Improving Plants through Genetic Engineering

Table 6.1 illustrates on the use of genetic engineering as a practical tool for the improvement of plants, considering them as chemical factories that use sunlight and atmospheric CO_2 as sources of energy and feedstock, respectively. As shown, plant transformation is performed with three main objectives: *(i)* incorporation of new traits, endowing plants with increased resistance to chemicals, pathogens or different stresses; *(ii)* the improvement of products naturally synthesized by plants through the increase of their quantity and/or quality; and *(iii)* the incorporation of genes to plants for the production of heterologous proteins or novel compounds with pharmaceutical or other industrial applications. We will consider examples and possibilities in each of these three categories outlined in Table 6.1.

Improving Plant Resistance to Chemicals, Pathogens and Stresses

Development of plants with higher resistance to many biotic and abiotic stresses is a main goal to be necessarily reached in order to solve the serious increased demand for food production, which should be performed with a maximal of land and water use economy (Somerville and Briscoe, 2001). Plant genetic transformation in this field is a key tool to produce economically relevant species with improved resistance to chemicals, pathogens and other stresses.

Production of plants with resistance to herbicides is a common example of plant transformation by genetic engineering for the yield increase of many crops (Collins and Shepherd, 1996; Willmitzer, 1999). Herbicides are chemicals that act inhibiting target enzymes involved in photosynthesis or the metabolism of amino acids essential to plants. The widely utilized herbicide glyphosate is an analog of phosphoenolpyruvate, a metabolite occupying a key central position in plant metabolism, including essential amino acid biosynthesis.

Table 6.1. Realities and potentialities in plant genetic transformation.

Plant transformation (main objective)	Possibilities	Examples	References
1. As a tool for basic research in plant biochemistry/ physiology	Expression, over- or under-expression or deletion of specific genes coding selected proteins (enzymes, translocators, hormones, receptors, etc.).	Studies on carbon allocation and partitioning. Characterization of enzymes and hormones involved in plant development. Studies on cellular signals for plant-microbe interaction	Stitt, 1994; Theologis, 1994; Mol *et al.*, 1995; Staskawicz *et al.*, 1995; Herbers and Sonnewald, 1996; Stitt, 1999
2. As a tool for the improvement of plant uses	a. Introduction of new traits to plants for increasing resistance to chemicals, pathogens, and abiotic stress	Transgenic plants with resistance to Glyphosate, pests, osmotic and water stress	Collins and Shepherd, 1996
	b. Manipulation to improve the quantity and/or quality of natural products derived from plants	Transgenic tomato with longer post-harvest life. Transgenic plants producing higher amounts or structurally modified fats, carbohydrates, proteins, vitamins. Plant transformation to improve production of secondary metabolites	Visser and Jacobsen, 1993; Collins and Sheperd, 1996; Sivak and Preiss, 1998; Heyer *et al.*, 1999; Morgan *et al.*, 1999; Ohlrogge, 1999; Della Penna, 2001; Somerville and Bonetta, 2001
	c. Production of heterologous proteins for pharmaceuticals and other industries	Transgenic plants producing cholera toxins and human enzymes and antibodies.	Collins and Shepherd, 1996; Lundblad and Kingdon, 1999; Fischer *et al.*,1999a-c.

Glyphosate inhibits EPSP (5-enolpyruvylshikimate-3-phosphate) synthase, an enzyme involved in the anabolic pathway of aromatic amino acids (Herrmann and Weaver, 1999). The effect of glyphosate on plants is non-selective and of broad spectrum; thus this compound is toxic not only for weeds but also for crops, by affecting protein synthesis (as a consequence of the effect on aromatic amino acids anabolism), and cell growth. In transgenic plants, tolerance to the herbicide is achieved by over-expression of the normal EPSP synthase, or by expression of a glyphosate-insensitive enzyme, naturally occurring in *Agrobacterium tumefaciens*. Alternatively, transformation includes the expression in plants of a bacterial glyphosate oxido-reductase, an enzyme that degrades the herbicide to non-toxic compounds (Herrmann and Weaver, 1999).

The increase in plant resistance to a number of pests has been obtained by transforming different crops with genes encoding for endotoxins from *Bacillus thuringiensis* (*B.t.*) (Collins and Shepherd, 1996). This aerobic, Gram-positive soil bacterium, produces a number of insect toxins. More important are δ-endotoxins (protein crystals formed during sporulation), with activity against a number of caterpillars. This approach represents a good alternative to chemical insecticides for controling many species of pests insects (McGaughey and Whalon, 1992; Della Penna, 2001; Somerville and Bonetta, 2001). A strategy followed for successful plant transformation, comprises the introduction of the *B.t.* toxin genes linked to a constitutive promoter allowing the expression of toxic proteins in all plant tissues. This methodology has the advantages of being environmentally-safe and exhibiting high efficiency; although disadvantages include high cost of production and low persistence, as well as the appearance of resistant insects, and are worthy of mention (McGaughey and Whalon, 1992).

Plant transformation using genes encoding for proteinase inhibitors (proteins naturally produced by certain plants that inhibit proteinase action) was also successfully applied to improve resistance to pests. The accumulation of the expressed proteinase inhibitor becomes toxic for herbivorous insects thus resulting in an effective and broad spectrum pest control strategy. Transgenic plants over-expressing hydrolytic enzymes (i.e. chitinase) and exhibiting enhanced resistance to fungal infections have also been developed (Della Penna, 2001; Somerville and Bonetta, 2001).

Many environmental stresses, including drought, high salinity, temperature extremes, toxicity by metals and other pollutants, UV-B radiation; significantly affect crop productivity (Smirnoff, 1998). Different approaches have been proposed to apply genetic engineering to obtain economically-important plants with increased resistance to a certain stress. In this sense, a very promising field is the metabolic engineering of plants for osmotic stress resistance (Nuccio *et al.*, 1999; Bartels, 2001). Various osmoprotectants (or compatible solutes) were found in plants and bacteria. Different plant species synthesize certain osmotically compatible metabolites and accumulate them at different amounts. Remarkably, there are plants that produce sugar alcohols (mannitol, sorbitol), and these metabolites may represent a major product of photosynthesis together with sucrose and starch (Stoop *et al.*, 1996). One strategy for plant transformation is to express genes in species that do not produce these compounds, or to accurately manipulate the level (and the time) of expression in those plants that normally synthesize them. The convenient engineering of this biochemical trait may allow not only the production of plants with increased growth capacity under osmotic stress conditions, but also those with improved resistance to other abiotic or biotic stresses (Bartels, 2001). Interestingly, it has been shown that some osmoprotectants are also useful to the plant for coping with oxidant conditions as well as with pathogen attack (Stoop *et al.*, 1996; Smirnoff, 1998).

Improving Quality and Quantity of Plant Products

Transgenic tomatoes, in which fruit ripening was conveniently modified (Flavr Savr tomato), was the first genetically engineered whole food sold commercially (Collins and Shepherd, 1996). The resultant fruit exhibits a longer postharvest life. This allows not only for a better shipping and handling of commercial goods, but also for the performance of complete fruit ripening on the plant, which yields a product with higher quality (improved flavor). Transformation of tomato plants was performed utilizing antisense technology to reduce the expression of polygalacturonase (PGase, the enzyme degrading pectin in the cell wall) by near 90%. PGase antisense gene was constructed by fusing a cDNA clone of the enzyme gene in reverse orientation to a constitutive promoter, and then introducing it into tomato plants. During gene transcription, transgenic cells produce normal as well as antisense (complementary) mRNA molecules. The latter interact by binding to normal mRNA molecules thus inhibiting translation into PGase protein. An alternative approach is being evaluated for the improved manipulation of fruit ripening by regulating synthesis of ethylene (a phytohormone involved in modulating biochemical processes related with ripening), with enzymes producing or degrading aminocyclopropane carboxylate (ACC synthase and deaminase, respectively) key targets for this purpose (Collins and Shepherd, 1996).

Engineering of plant fatty acid composition is one of the most developed areas in the field of genetic transformation used to improve the quality of plant food products (Topfer *et al.*, 1995; Collins and Shepherd, 1996; Ohlrogge, 1999; Somerville and Bonetta, 2001). A main objective in these studies, is to modify chain length and degree of saturation of the fatty acids present in triacylglycerols, the plant storage oils. Successful results were obtained by manipulation of the level of expression (over- or under-expression) of genes coding for specific thioesterases (which catalyze the hydrolytic release of fatty acids bound to acyl-carrier protein) or desaturases (which specifically introduce unsaturation in the fatty acid chain). Derived transgenic plants synthesize modified or novel oil compounds having relevance for the production of healthier food as well as for their application in industrial processes (with the added advantage of constituting a renewable, biodegradable, resource) (Collins and Shepherd, 1996). Attempts to modify oilseed fatty acid composition by genetic engineering to produce other chemical structures (i.e. conjugated double bonds, epoxy functions) are now being carried out (Ohlrogge, 1999).

Manipulation of the content and composition of carbohydrates is a primary challenge for plant genetic engineering (Collins and Shepherd, 1996; Della Penna, 2001). It is worth remembering that carbohydrates (mainly starch and sucrose) are the major final photosynthetic products. Ultimately, an improvement in the process of inorganic carbon assimilation by plants may be attained straightforwardly so as to increase the total biomass available on the earth (Walker, 1995). Although definite success with respect to modifying the whole process is to date far from being reached (mostly because of the complexity of metabolism and compartmentation existing in plants, which will be discussed later in this chapter), important partial goals have been already claimed in terms of understanding of the carbohydrate allocation and partitioning in plants (Stitt, 1994; Collins and Shepherd, 1996).

Starch is the main storage polysaccharide found in plants. Its biosynthesis plays different functions during plant growth and development and critically determines the yield of many crops (corn, wheat, rice, potato, yam, cassava). Starch has multiple uses, as it constitutes the main source of food for humans and is utilized in a wide number of industrial processes (Sivak and Preiss, 1998). Polysaccharide synthesis takes place in the chloroplast (in photosynthetic cells) or in the amyloplast (in storage tissues), and it involves three enzymes: ADPglucose pyrophosphorylase (ADPGlcPPase), starch synthase, and branching enzyme. ADPGlcPPase catalyzes the production of ADPglucose (ADPGlc) from glucose-1-P and ATP, and is the key regulatory step in the storage polysaccharides biosynthetic pathway. Because of its role in starch synthesis, ADPGlcPPase is given as an example for the understanding of control and regulation of a metabolic flux (see below). ADPGlc is the glucosyl donor molecule utilized by starch synthase to elongate an α-1,4-glucan chain. Then, branching enzyme introduces α-1,6-branches. The resulting product (starch) is a mixture of two polysaccharides: amylose (the linear α-1,4-glucan) and amylopectin (the branched polymer) (for further details see reviews by Ball, 1995; Iglesias and Podestá, 1996; and Sivak and Preiss, 1998).

A remarkable success in the use of plant biochemical engineering for the improvement of food quality traits was the production of transgenic potatoes accumulating increased amounts of starch (Collins and Shepherd, 1996; Sivak and Preiss, 1998; Della Penna 2001). In this example, the strategy followed was to express a mutant (unregulated) ADPGlcPPase from *E. coli* in potato plants, with a transit peptide sequence that targeted it to the stroma of amyloplasts. The use of a mutated non-regulated enzyme circumvented the necessity for the presence of specific metabolites that allosterically regulate ADPGlcPPase activity (see below). By this procedure the synthesis of ADPGlc within the amyloplasts was enhanced (constantly and independently of regulator metabolites) and this resulted in higher amounts of starch (Collins and Shepherd, 1996; Sivak and Preiss, 1998). The transgenic product possesses a series of advantages, e.g. it is more suitable for storage and gives healthier fried food derivatives. By similar procedures, high starch tomato and canola plants were obtained (Collins and Shepherd, 1996).

The success in the manipulation of the amount of starch synthesized by plants strongly supports the prospect of important achievable aims in the field of biodegradable polymers. The possibility of modifying the quality of starch would be a highly relevant result and can be visualized as a necessary next step to be reached in plant biotechnology. The relative content of amylose and amylopectin greatly influences the physicochemical properties of starch and this determines its use in different industrial processes (Visser and Jacobsen, 1993; Heyer *et al.*, 1999). Approaches to modify the ratio of polysaccharides includes manipulation of starch synthase isoenzymes and branching protein isoforms, as well as the proper ADPGlcPPase, after it was shown that alterations in the supply of ADPGlc can affect the production of amylose (Visser and Jacobsen, 1993; Sivak and Preiss, 1998; Heyer *et al.*, 1999; Lloyd *et al.*, 1999). Another critical factor affecting the quality of starch for industrial uses is the degree of phosphorylation of the polysaccharide molecule. The gene responsible for the phosphorylation of glucans has been recently cloned and utilized to genetically engineer potato starch (Heyer, 1999). All these developments open new possibilities for a great improvement in production of modified polymeric carbohydrates in the near future.

Plant secondary metabolites are a series of phytochemicals including alkaloids, polyphenols, steroids, terpenoids, anthocyanins and anthraquinones. They are specific to special groups of plants and, in some cases, are only produced in certain tissues of a given species. These compounds have a high commercial value because of their extensive industrial use mainly as drugs, perfumes, pigments, agrochemicals and food flavors. Good examples of valuable, currently useful pharmaceuticals are the anticancer drugs vinblastine and taxol (Morgan *et al.*, 1999). Their market relevance as chemicals, as well as their key roles in plant defence against pathogen attack, make it obvious that secondary metabolism is a major target for plant transformation. Efforts are being made to use MCE to improve the performance of whole biochemical pathways and thereby to attain improvements in the *in vivo* production of these compounds (usually synthesized in quite low amounts). Important advances have been reached in the development of plant cell culture techniques for the production of phytochemicals, as well as in the transformation of certain species to improve natural defence response to pathogens (Collins and Shepherd, 1996; Dixon *et al.*, 1996; Morgan *et al.*, 1999).

Plant proteins are deficient in certain amino acids that are essential for humans and animals, resulting in a shortfall in the nutritional quality of foods derived therefrom. Typically, cereal proteins are poor in lysine and tryptophan, whereas legume and vegetables contain low amounts of methionine and cysteine. Genetic engineering emerges as a very convenient tool, with remarkable advantages over traditional breeding regimes, to overcome this deficiency. Identification of seed storage proteins rich in sulfur amino acids in plant members of the Brazil nut family, followed by the engineering of their expression in target plants, are important approaches to the goal of obtaining methionine-rich transgenic seeds (Collins and Shepherd, 1996).

Worth of mention are also the goals reached through genetic manipulation to improve the content of certain vitamins in plants (Della Penna, 2001). One example is the expression of the enzyme γ-tocopherol methyltransferase in *Arabidopsis* seeds to significantly increase the amount of vitamin E; which represents a very promising strategy to manipulate levels of this relevant antioxidant (vitamin E) in other plant seeds (such as soybean, maize, canola), to obtain oils with improved quality for the human diet (Shintani and Della Penna, 1998). A second example is the remarkable product called "golden rice" after the yellow endosperm exhibited by rice that resulted from engineering plants for the simultaneous expression of three enzymes (two from plants and one from bacteria) involved in the carotenoid biosynthetic pathway (Ye *et al.*, 2000). The final product (golden rice) contains high levels of β-carotene (precursor of vitamin A), thus representing a very important tool to cope with a serious vitamin deficiency in the developing world (this product having the additional strategic advantage that rice is a major commodity in developing countries).

Using Plant Genetic Engineering to Produce Heterologous Proteins

Plants represent a very convenient biological system for the production of recombinant proteins, a fact increasingly used for medicinal applications (ranging from antibodies to diagnostic proteins) as well as in other industries [i.e. the production of fibers exhibiting properties of silk, collagen or keratin; with the expression of elastin-like polymer in plants being a recent application in this way (see Guda *et al.*, 2000)]. Recent reviews in Biotechnology and Applied Biochemistry (Lundblad and Kingdon, 1999; Fischer *et al.*, 1999a-c) clearly highlight the relevance of a "molecular farming" strategy for the production of eukaryotic proteins. Plants offer reduced technical, ethical, safety issues, and costs, as compared with the use of mammalian-cell cultures or transgenic animals. Genetically engineered plants can synthesize fully-folded functional proteins (even complex, glycosylated proteins), since they possess similar post-translational and co-translational modification mechanisms to those of animal cells. In addition, the photosynthetic capacity of plants, combined with the advances of modern agriculture, enable low-cost production of large amounts of recombinant macromolecules. The possibility of expression of the heterologous protein specifically in a certain tissue, e.g. seeds, is a plus, as it improves the capacity for storage of active biomolecules. Successful achievements worthy of mention include the expression in plants of cholera toxin subunits, human enzymes and antibodies (Collins and Shepherd, 1996).

Tools for the Manipulation and Transformation of Plants

Specific methods have been developed for manipulating and introducing foreign DNA into plant cells. We will briefly consider these methods that are the necessary tools for plant genetic engineering: *(i) in vitro* plant tissue culture techniques, and *(ii)* plant transformation systems.

(i) In vitro *plant tissue culture*. Tissue culture of higher organisms is based on the theory developed in 1838 by Schwan and Schleiden, establishing that, because of its totipotency, a single plant cell is self-sufficient and able to generate a whole plant (Pierik, 1988). In 1939, Nobécourt, Gautheret and White were the first to successfully obtain plant tissue culture, by growing tomato root tips using an artificial medium. It was after the discovery of phytohormones and their action as growth phytoregulators, that appropriate techniques for the culture of plant cells were rapidly developed (Pierik, 1988).

Plants can be grown *in vitro* under sterile culture conditions in a medium containing a carbon source, minerals, growth factors, and regulators (Dixon, 1985; Pierik, 1988). Different types of culture methods comprise the use of single cells or part of a plant, to initiate growth. The latter has been extensively utilized to develop techniques of micropropagation, i.e. generation of plants by asexual means under laboratory conditions. This technology is being successfully applied for different commercial plants, including ornamentals (orchids, ferns), woody species (*Eucalyptus*, *Pinus*, *Sequoia*), and several crops (banana, carrot, cassava, celery). The culture of single cells is mainly utilized for the production of secondary metabolites under controlled conditions.

Intact whole plants can be cultured from seeds that are aseptically germinated or from pieces of different tissues. A variant of this approach is the culture of embryos (or embryo-like structures) previously isolated from the seeds or obtained by somatic embriogenesis from other plant tissues or organs. The explant (i.e. the initial piece of the plant introduced *in vitro*) is conveniently selected from meristems, shoot tips, axillary buds, leaves, roots, stems, petioles, or flower parts; then disinfected (to clean out of contaminating surface microorganisms) and cultured in defined medium to promote morphogenesis. Figure 6.1 shows that the generation of a plant can be made via organogenesis (direct morphogenesis into shoots and roots) or through somatic embryogenesis (somatic embryos are induced prior to morphogenesis). Moreover, both procedures, morphogenesis and organogenesis, can be achieved directly or indirectly, via the initial transformation of a sterile piece of tissue into a callus (a mass of unorganized, undifferentiated cells) or suspension culture (Fig. 6.1).

On the other hand, single cells are obtained enzymatically or mechanically from a callus of tissue culture in suspension, and then maintained in continuous growing in aerated liquid media. An interesting variant is the culture of plant protoplasts (plant cells without the walls), which can be prepared from cell suspensions after incubation in media containing mannitol at hypertonic levels (to detach the cytoplasmic membrane from the cell wall) followed by enzymatic digestion of the wall with pectinase.

(ii) Plan transformation systems. The development of systems for the successful genetic transformation of plants needs the consideration of distinct limitations and problems; this implying the utilization of specific methods (Owen and Pen, 1996; Birch, 1997). Thus, the cell wall is a primary barrier to the introduction of foreign DNA into plant cells. In addition to this problem, common to all plants, there are difficulties inherent in the differences existing between plant tissues and species that determine unequal responses to a transformation protocol. As a whole, the methods that have been developed for plant genetic transformation can be conveniently divided into vector-mediated or vector-free systems according to the way the genetic material is transferred.

Vector-mediated gene transfer systems for plants, are based in the use of bacteria or viruses. Two species of the genus *Agrobacterium*: *A. tumefaciens* and *A. rhizogenes*, are the main bacterial vectors for plant transformation (Tepfer, 1990; Hooykaas and Schilperoort, 1992; Zupan and Zambryski, 1995). *A. tumefaciens*, a Gram-negative soil bacterium causing a tumour (crown gall disease) in leguminous dicot plants, is the best known and more commonly utilized vector. Bacterial infection initiates at sites of mechanical wounding in the plant through chemotaxis, induced by phenolic compounds released from damaged cells. Plant infection and disease formation by *Agrobacterium* involve gene transfer between cells, resulting in the integration of a DNA sequence from the invading pathogenic bacterium into the DNA of the host plant cell. The latter induces cell proliferation with unusual biochemical characteristics, as they produce opines (amino acids utilized by *Agrobacterium* as a source of carbon and nitrogen) and exhibit hormone-independence (auxins and cytokinins) for growth.

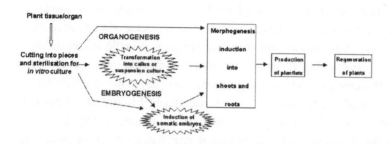

Figure 6.1. Different strategies for plant tissue culture. Tissues can be cultured to generate shoots and roots (organogenesis) or somatic embryos (embryogenesis). In both cases the procedure can be made directly or indirectly via the previous transformation into callus or suspension culture.

The genetic material (named T-DNA, from "Transferred **DNA**") causing the tumour is carried on a 200 kbp plasmid (the **Ti** plasmid from "Inducing-inducing") that is separate from the *Agrobacterium* main chromosome. During infection, T-DNA is transferred from the bacterium into plant cells, where it enters the nucleus and integrates into the plant genome. In addition to the T-DNA (T-region), Ti plasmid also has the virulence region (coding for proteins involved in the T-DNA transfer) and the opine catabolic region (allowing the use of opines by bacteria).

Transformation systems for plant genetic engineering using *Agrobacterium,* take advantage of the fact that deletion of genes of the virulence and T-region, from the Ti plasmid, produces no adverse effect on the transfer and integration of T-DNA. Thus, with different strategies, Ti plasmid is engineered to construct the transformant vector by deleting the coding region of the opine biosynthetic genes and addition of the foreign genes. Transformation using Ti plasmid as a vector is useful for plants susceptible to infection by *Agrobacterium* (mostly dicotyledonous species) and is generally applied to cultures of callus or leaf discs, where cut edges provide the wound region to initiate the bacterial infection (Owen and Pen, 1996).

Certain plant viruses have been utilized as vectors to transform different species. Most of them are single-stranded RNA viruses and are of relevance because of their potential for high levels of expression of transgenes; a wide range of plant species are susceptible to virus infection (Owen and Pen, 1996). The virus-mediated plant transformation system is particularly important for the production of vaccine epitopes (e.g. the use of tobacco mosaic tobamovirus, cowpea mosaic comovirus and johnsongrass mosaic potyvirus, Collins and Shepherd, 1996).

Vector-free systems for the transfer of naked-DNA into plant cells, comprise the use of chemical, physical (mechanical), and electrical methods (Owen and Pen, 1996). In the three sorts of methods, the transfer is preferably performed on protoplasts, in order to overcome the barrier of the plant cell wall. Chemical methods include techniques using polyethyleneglycol to facilitate the uptake of DNA by cells or to fuse liposomes (encapsulating the DNA) with protoplast (Krens *et al.*, 1982; Gad *et al.*, 1990). Among mechanical methods, it is worth considering DNA microinjection into the nucleus of individual cells through fine glass needles (Neuhaus and Spangenberg, 1990); the formation of long needle-like crystals (whiskers) of silicon carbide which penetrate the cells allowing the entry of DNA (Kaeppler *et al.*, 1990); and the bombardment of cells with microprojectiles of tungsten or gold coated with DNA (Christou, 1992). Finally, electroporation is based on the increased permeability of the plasma membrane to hydrophilic molecules on application of well-defined electrical pulses to protoplasts or cells suspensions containing the transformant DNA (Owen and Pen, 1996).

Plant Metabolism: Matter and Energy Flows and the Prospects of MCA

Despite the many relevant goals achieved in plant genetic engineering (most of them previously described in this chapter), the understanding of the flux of matter and energy occurring in plant cells is far from being complete; and research on MCA in plants is notably scarce. It can be stated that current knowledge of the control of plant metabolism is mostly qualitative; with the development of studies allowing an analysis at a quantitative level being in its infancy. This is an important challenge for plant scientists, since a significant advance should be gained in the area in the next few years as a prerequisite for the rational improvement of tools and designs in MCE as applied to plant systems.

A number of features make plants outstanding complex organisms in which to perform studies on MCA. Plants are able to grow and reproduce under quite a diverse range of environmental conditions. This property is based on the occurrence of a flexible metabolism (represented by multiple, inter- and trans-connected metabolic pathways), and a variety of regulatory processes that account for the complex physiological and biochemical scenario of plant cells (apRees and Hill, 1994). On the other hand, plant metabolism is highly compartmented, with many pathways, completely or partially, duplicated in different compartments.

Metabolic Compartmentation in Plant Cells

In addition to nucleus, mitochondria and other organelles found in fungi and animals, cells of photosynthetic eukaryotes possess plastids. These are specific self-replicating organelles surrounded by an envelope (formed by two membranes) and occurring in a variety of types, sizes, shapes and colors. The different plastids exhibit distinctive metabolic capacities: typically chloroplasts contain chlorophyll pigments and are the site of photosynthesis; or amyloplasts, found in non-green tissues and specialized in the storage of starch (Newcomb, 1990). A good example of the complexity introduced by metabolic compartmentation is that of glycolysis, which in plants involves one set of enzymes in the cytosol and another in the plastid, operating simultaneously and largely independently (Plaxton, 1996). The extensive compartmentation and duplication of metabolic pathways in plants makes difficult studies of flux analysis. According to the quantitative methodology utilized, metabolite concentrations and the measurement of *in situ* enzyme activities (which should correspond to the level of metabolite and to the activity of the isoenzyme in each specific intracellular compartment) may be needed (apRees and Hill, 1994).

An example of the complex network of reactions operating in the different intracellular compartments within a photosynthetic cell is shown in Fig. 6.2. The figure illustrates the interaction between phosphate uptake, carbon assimilation, and respiration, in the unicellular alga *Selenastrum minutum*. Interestingly, green algae (especially those unicellular species) are highly amenable organisms for studying and quantifying metabolic fluxes in photosynthetic cells. Thus, because of its phylogenetic and genetic properties, the species *Chlamydomonas reinhardtii* has been called "green yeast", to emphasize the potential relevance of this organism for biological studies (Goodenough, 1992).

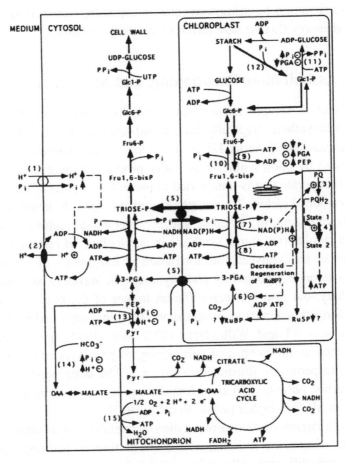

Figure 6.2. Schematic model for the flux of carbon in the green alga *Selenastrum minutum*. The model explains interactions between Pi assimilation, photosynthesis and respiration in a carbon assimilating (autotrophic) cell. Taken from Gauthier and Turpin, 1997, by permission of Blackwell Science.

Recently, it was reported genetic manipulation of the microalga *Phaeodactylum tricornutum* to modify its throphic capacity (Zaslavskaia *et al.*, 2001). The alga was genetically engineered by the introduction of a gene encoding a glucose transporter resulting in cells than can thrive on exogenous glucose in the absence of light. The work by Zaslavskaia *et al.* (2001) worthily demonstrated that the introduction of a single gene in an organism could produce substantial changes in its metabolism (remarkably in a complex metabolism of the type exemplified in Fig. 6.2). The advantage of the modified microalga is its dual capacity to grow photo- or hetero-trophically and thus being useful for the efficient exploitation by using fermentation technology as well as for studies of carbon and energy fluxes under different throphic conditions.

Carbon Assimilation, Partitioning, and Allocation

Another level of complexity in plant metabolism arises from the existence of distinct photosynthetic (typically leaves) and heterotrophic (i.e. roots and seeds) tissues (Iglesias and Podestá, 1996). In the leaves, atmospheric CO_2 is fixed into carbohydrates by photosynthesis. A major product of the process, sucrose, is a mobile carbohydrate that is transported to the different parts of the plant where it provides carbon skeletons for non-photosynthetic cells. In heterotrophic tissues, sucrose is allocated to different specific metabolic pathways, occurring in distinctive plastids (i.e. starch synthesis in amyloplasts of reserve tissues such as endosperm, or triacylglycerol accumulation in leucoplasts of certain seeds). In this way, carbon assimilated by photosynthesis in leaves is partitioned not only intracellularly, but also between different tissues of the plant. Thus, interactive source-sink relationships for carbon metabolism are established between different tissues (Sonnewald and Willmitzer, 1992; Stitt, 1994; Iglesias and Podestá, 1996). Carbon assimilation, and partitioning of photosynthate between the leaf and the endosperm, are illustrated in Fig. 6.3. The understanding of the whole process: carbon photoassimilation, partitioning, and utilization of photosynthate, is critical for MCE in higher plants (Beachy, 1997).

The high degree of compartmentation of metabolism in plants includes the distribution of enzyme activities and sequestration of pathways (or a part of them) between different cells and/or subcellular compartments; as well as the intra- and inter-cellular transfer of different metabolites. As pointed out by Emes *et al.* (1999), nowhere is the above better exemplified than in the area of carbohydrate metabolism. We show the complexity of carbohydrate synthesis and utilization in source and sink tissues of plants in Fig. 6.3 A and B, respectively.

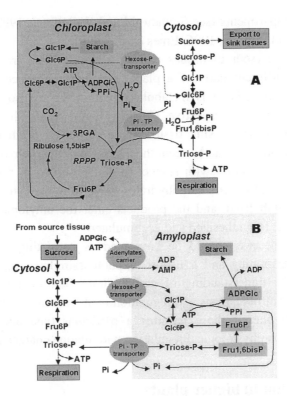

Figure 6.3. Diagram of carbon flux in cells of (A) photosynthetic (source) (B) and non-photosynthetic (sink) tissues of higher plants. In A the pathway for carbon fixation and partition between sucrose and starch in a leaf cell is shown. In B, it is schematized the pathway conducting to the final allocation of photoassimilates in a tissue accumulating starch (such as endosperm). Exchange of metabolites between cytosol and plastids are described with the involvement of specific translocators: the hexose-phosphate (Hexose-P) transporter, the Pi-triose-phosphate (Pi-TP) transporter, and the adenylates carrier.

As shown (Fig. 6.3A), carbon assimilation in the photosynthetic cell occurs in the chloroplast, through a cyclic pathway known as RPPP (from **R**eductive **P**entose **P**hosphate **P**athway) (Iglesias *et al.*, 1997). For the sake of clarity, the series of reactions occurring in the RPPP are shown separately in Fig. 6.4. From intermediate metabolites of this cycle, the respective routes leading to the two major products of photosynthesis: starch and sucrose, are initiated (Fig. 6.3A). In this way, photoassimilates are mainly partitioned between chloroplastic starch (the polysaccharide serving for the temporary accumulation of carbohydrates in leaves) and cytosolic sucrose (the disaccharide delivered to other non-photosynthetic parts of the plant). As illustrated also in Fig. 6.3A, major metabolites transported across the chloroplast envelope are triose-P; a process involving a specific translocator that interchanges Pi and triose-P between intracellular compartments (see Sonnewald and Willmitzer, 1992; Iglesias *et al.*, 1993; Iglesias and Podestá, 1996; Emes *et al.*, 1999).

Photosynthetic (source) tissues export sucrose, which travels through the phloem to the different parts of a plant (sink tissues). Once sucrose reaches heterotrophic plant cells, it is metabolised within the cytosol to the different routes occurring in the respective plastids. Figure 6.3B shows the pathway followed in endosperm, a sink tissue where carbon is accumulated as starch inside the amyloplast. As can be seen, the major flux of carbohydrates in this case is in the opposite direction with respect to that observed in the photosynthetic cell. Thus, in the endosperm, carbohydrates move from the cytosol to the plastid, as sucrose is metabolised, and the products enter the amyloplast, where starch (the final product) is synthesized and accumulated.

The overall process should be envisaged as consisting of metabolism in source cells, partitioning and transport of photoassimilates, and metabolism in sink tissues. The understanding of carbohydrate flux in a whole plant thus requires a flux control analysis and regulation in at least three different metabolic processes. Each one of these requires a particular type and distribution of enzymes as well as physiological regulation with unique characteristics.

Carbon fixation in higher plants

Eventhough complex at first sight, the scheme shown in Fig. 6.3A is an oversimplification of carbon assimilation and partitioning in photosynthetic cells from higher plants. A more detailed picture needs to consider flux fate as well as additional reactions of the carbon assimilation process. The reaction of atmospheric CO_2 fixation in the cyclic RPPP is catalyzed by ribulose-1,5-bisP carboxylase/oxygenase (abbrev. as Rubisco) (Fig. 6.4). The result of carboxylation of ribulose-1,5-bisP is the production of 2 molecules of 3P-glycerate (3PGA). This metabolite is then converted, through the RPPP, to triose-P [glyceraldehyde-3P (Ga3P) and dihydroxyacetone-P (DHAP)] in two consecutive reactions that consume ATP and NADPH (produced by photosynthetic electron transport at the thylakoid membrane). The cycle continues in a series of reactions that interconvert compounds of three to seven carbon atoms and that regenerate the initial acceptor of CO_2 (ribulose-1,5-bisP) with the additional consume of ATP (Fig. 6.4). Thus, the RPPP is autocatalytic and allow assimilating one inorganic carbon molecule into one organic metabolite per turn. Figure 6.4 shows that to produce net synthesis of one hexose-P molecule, the RPPP needs to fix 6 CO_2, utilizing 18 ATP and 12 NADPH molecules.

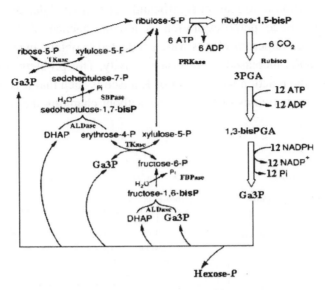

Figure 6.4. The reductive pentose phosphate pathway (RPPP), showing the net production of one hexose-P from the photoassimilation of six CO_2 molecules. The non-reversible reactions catalyzed by Rubisco, fructose-1,6-bisP phosphatase (FBPase), seudoheptulose-1,7-bisP phosphatase (SBPase), and P-ribulokinase (PRKase); as well as the reversible steps mediated by transketolase (TKase) and aldolase (ALDase) are marked with the name of the enzymes. Specific abbreviations for the figure are: DHAP, dihydroxyacetone-P; GAP, glyceraldehyde-3P.

As its name indicates, Rubisco is able to catalyze carboxylation as well as oxygenation of ribulose-1,5-bisP:

ribulose-1,5-bisP + CO_2 → 2, 3P-glycerate

ribulose-1,5-bisP + O_2 → 3P-glycerate + 2P-glycolate

Both reactions take place within the chloroplast of photosynthetic cells, the occurrence of each being mainly determined by the ratio of substrates (CO_2:O_2) within the plastid (Iglesias *et al.*, 1997). The oxygenase activity of Rubisco initiates a metabolic pathway known as photorespiration, whose products spread among different intracellular compartments (chloroplast, cytosol, peroxysome, mitochondria) (Fig. 6.5) (Iglesias *et al.*, 1997; Emes *et al.*, 1999). As shown in Fig. 6.5, photorespiration consumes O_2 and releases CO_2, clearly being an opposing pathway to carbon assimilation.

Thus, photorespiration not only adds a degree of complexity to carbon metabolism and partitioning in photosynthetic cells, but also affects the yield of the process of carbon fixation in higher plants. Under certain environmental conditions, photorespiration can strongly affect the efficiency of photosynthesis. In example, at high temperatures, O_2 solubility decreases but relatively much less than that of CO_2, thus resulting in a lower CO_2:O_2 intracellular ratio. Under these conditions, net carbon assimilation significantly decreases (Iglesias *et al.*, 1997; Edwards, 1999). Since photorespiration is a consequence of the oxygenase activity of Rubisco, numerous efforts were conducted to obtain an enzyme having only (or substantially increased) carboxylase activity. This problem is unsolved at present, because the reaction mechanism of Rubisco is such that it is not possible to affect one of the activities of the enzyme without changing the other (Iglesias *et al.*, 1997).

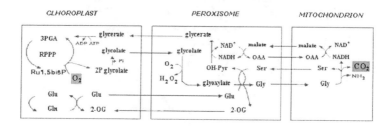

Figure 6.5. Photorespiration, the pathway followed by carbon between different compartments of a photosynthetic cell (chloroplast, peroxisome and mitochondrion) as a consequence of the oxygenase activity of Rubisco. Specific abbreviations for the Figure are: 2-OG, 2-oxoglutarate; Gln, glutamine; Glu, glutamate; Gly, glycine; OH-Pyr, hydroxy-pyruvate; Ru1,5bisP, ribulose-1,5-bisP; Ser, serine.

During evolution the path followed to decrease or eliminate photorespiration was different, and comprised the addition of a series of metabolic steps to the process of photosynthetic CO_2 fixation. The RPPP cycle is common to all autotrophic cells of higher plants, being the obligate primary route for carbon assimilation. Because the first product of CO_2 fixation via the RPPP is a 3-carbon compound (3P-glycerate, see Fig. 6.4), this route is also known as the C_3 cycle. C_3-plantsis the name given to those species possessing only this route for carbon assimilation. Certain plants have an additional (not alternative) pathway that operates separately, spatially or temporally, with RPPP. In such additional route, the reaction of atmospheric CO_2 fixation is catalyzed by phosphoenolpyruvate carboxylase (PEPCase) an enzyme that cannot use O_2 as a substrate:

$$PEP + HCO_3^- \rightarrow oxaloacetate + Pi.$$

This reaction starts a cyclic metabolic pathway named the C_4 route, because of the 4-carbon compounds (oxaloacetate, malate, aspartate) produced during CO_2 fixation. In this cycle, oxaloacetate is reduced to malate or transaminated to aspartate and, after a series of steps a 4-carbon metabolite is decarboxylated at the chloroplast. The CO_2 released serves as a substrate for Rubisco that ultimately refixes it via the RPPP. The C_4 route is cyclic because the C_3-carbon compound resulting from the decarboxylation is then utilized to regenerate PEP, the substrate of PEPCase.

Figure 6.6 shows the operation of the C_4 route in a C_4-plant, where the whole carbon assimilation process takes place with the involvement of two different photosynthetic cells: mesophyllic and bundle-sheath cells. C_4-plants are characterized by exhibiting a particular cellular anatomy in their photosynthetic tissues that is of key physiological relevance for carbon assimilation (Iglesias *et al.*, 1997; Edwards, 1999). As shown (Fig. 6.6), fixation of atmospheric CO_2 by PEPCase occurs in the cytosol of mesophyll cells. The C_4-carbon product is metabolised in the chloroplast or cytosol of the same cell, then transported to the bundle-sheath cell where, after being delivered to the chloroplast, it is decarboxylated. There, the RPPP is operative for the assimilation of CO_2. The C_3-metabolite produced by decarboxylation is recycled as it is metabolised with the final regeneration of PEP, a reaction occurring at the chloroplast of mesophyllic cells.

In Fig. 6.6 it can be visualized that in a C_4-plant, the additional C_4-pathway physiologically operates as a biochemical device pumping CO_2 from the atmosphere to the cellular place where the RPPP is localized. In this way, the $CO_2:O_2$ ratio at the chloroplast of bundle-sheath cells increases, thus favoring the occurrence of the carboxylase (over the oxygenase) activity of Rubisco. Consequently, under certain environmental conditions, C_4-plants exhibit a better performance in terms of photosynthetic efficiency than C_3-species; mostly because in the former photorespiration is markedly diminished or eliminated. Species of C_4-plants include the crops maize, sugar cane, sorghum and millet (Iglesias *et al.*, 1997; Edwards, 1999).

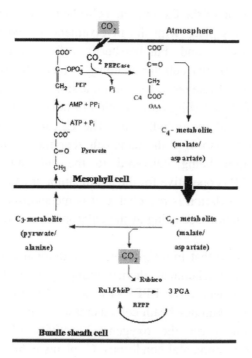

Figure 6.6. The carbon flux in a C_4-plant, involving two different photosynthetic cells: mesophyll and bundle sheath cells. C_3 and C_4 indicate 3- or 4-carbon compounds involved in the metabolic pathway.

A variant of this mechanism is found in plants where the C_4 pathway and the C_3 cycle function both in a single photosynthetic cell, but operate during different time periods. These plants usually grow in desert regions and are named CAM-plants (CAM after **C**rassulacean **A**cid **M**etabolism) because the first species characterized as performing this metabolism belonged to the Crassulaceae family (Iglesias *et al.*, 1997). In CAM-plants, atmospheric CO_2 is fixed during the night via PEPCase. The oxaloacetate thus produced is reduced to malate, which accumulates in the vacuole to attain very high levels (from this derives the term Acid in the pathway's name). During the day, malate is delivered to the chloroplast, where it is decarboxylated, the CO_2 released is refixed by Rubisco and finally assimilated via the RPPP. In this way, CAM-plants not only concentrate CO_2 in the chloroplast during the day (increasing the $CO_2:O_2$ ratio and thus reducing photorespiration); but also exchange gases with the environment only during the night (which avoids excessive loss of water). The latter is critical for organisms coping with extreme conditions existing in the desert, since the water economy has to be maximized for a better survival (Iglesias *et al.*, 1997; Edwards, 1999).

Based on strategies followed by nature to "metabolically engineer" plants during evolution to optimize carbon assimilation under different environmental conditions, efforts are being performed to design modified plants that perform photosynthesis more efficiently. Thus, the possibility of converting C_3-plants of economic relevance (wheat, rice, potato, soybean, among others) into C_4-plants, would theoretically increase their productivity by reducing photorespiration. Similarly, and emulating CAM-plants, it should be possible to increase the cultivable area of the planet after fine-tuned genetic engineering of certain species allowing them to grow in more arid habitats. However, it has to be considered that carbon metabolism is very complex in plants. At this point it would be convenient to review Fig. 6.2 and to consider what needs to be integrated to it from Figs. 6.3, 6.4, 6.5, and 6.6. Furthermore, a coordination of reactions in time and space for the desired pattern of operation will also be necessary (see Chapter 1).

Despite this complexity, some initial results suggest that important goals could be obtained by engineering carbon assimilation in plants. Recently, Ku *et al.* (1999) successfully transformed (using an *Agrobacterium*-mediated transformation system) the C_3 crop rice by the introduction of the intact gene of maize PEPCase. Transgenic plants showed high-level expression of the maize gene, with high PEPCase activity in leaves (where the enzyme reached up to 12% of the total soluble protein). Interestingly, transgenic rice plants exhibited reduced O_2 inhibition of photosynthesis (Ku *et al.*, 1999), thus showing a potential improvement in carbon assimilation. From this, it seems possible that relatively simple changes in enzyme activities can result in the successful improvement of carbon assimilation after the effective operation of complementary metabolic pathways in a single photosynthetic cell (in other words: apparently, the operativeness of efficient carbon fixation is not strictly dependent on the existence of a particular anatomy of cells as occurs in C_4 plants). The recent discovery of one particular mechanism of C_4-photosynthesis in the aquatic angiosperm *Hydrilla verticillata* (characterized by the operation of the C_4 pathway during the day in only one photosynthetic cell) reinforces the above possibility, as it seems to be a natural species where effective C_4-photosynthesis takes place without necessity of a complex cellular anatomy (Edwards, 1999).

MCA Studies in Plants

Studies carried out by using the methodology known as reversed genetics are giving relevant information for the understanding of carbon flux and regulation in higher plants (Stitt, 1999). Reversed genetics combines metabolic biochemistry and molecular genetics by utilizing the strategy of obtaining transformed plants, where the expression of one specific enzyme is progressively decreased and/or increased. This can be done by classical genetics, by varying the number of functional gene copies; or by transforming plants with antisense or sense constructs of specific genes. Quantitative analysis of changes in metabolic fluxes is then employed. Through the use of this experimental approach, the role of different enzymes on the control of photoassimilation pathway flux, was determined. The analysis comprised the study of enzymes catalyzing irreversible (fine-regulated) as well as reversible (non-regulated) steps from the RPPP, and the pathways for starch and sucrose synthesis in plant cells.

Results showed many relevant effects on metabolic fluxes and their control in photosynthetic cells of higher plants (Stitt, 1999). It was determined that control of flux is usually distributed between several enzymes catalysing both, regulatory and non-regulatory steps. Additionally, it was observed that regulated enzymes are present in excess (regulation of activity provides compensation of metabolic variations); whereas non-regulated enzymes are not expressed in a large excess (rationally attributed to the necessity for avoiding severe limitation of flux under ambient conditions). On the other hand, it was shown that the contribution of certain enzymes to the control of flux depends on the short-term conditions under which the measurement is performed, as well as on the environmental conditions for plant growth on a long-term period basis (Stitt, 1999).

These principles are demonstrated by analysis of the contribution of enzymes of the RPPP to the control of flux of carbon assimilation. Thus, under certain conditions, decreases of 40% or more in the activity of finely regulated enzymes of the cycle (those catalysing irreversible reactions: namely Rubisco, fructose-1,6bisP phosphatase, sedoheptulose-1,7bisP phosphatase, and P-ribulokinase; see Fig. 6.4) lead to slight or no inhibition of photosynthesis.

On the other hand changes in the levels of non-regulated enzymes of the metabolic cycle involving aldolase and transketolase (see Fig. 6.4), considerably affect carbon photoassimilation under the same conditions (Stitt, 1999). It has also been reported that changes in Rubisco levels slightly affect (control coefficient of 0.1) or proportionally modify (control coefficient of 0.9) the rate of photosynthesis depending on whether light conditions are constantly moderate or suddenly increased from a low steady level, respectively.

Of interest are studies showing how compartmentation and the existence of branch points are important factors to be considered for MCA or MFA in higher plants. As an example, it has been pointed out that the different contribution of chloroplastic and cytosolic P-glucoisomerase (PGI, catalysing the reaction: Glc6P ⟺ Fru6P) to photoassimilates partitioning between starch and sucrose in a photosynthetic cell (Stitt, 1999). A revision of Fig. 6.3A is useful for highlighting the occurrence of PGI in two intracellular compartments of photosynthetic cells. In the chloroplast, starch is synthesized from fructose-6P (an intermediate metabolite of the RPPP) by reversible reactions catalyzed by PGI and P-glucomutase (PGM, catalysing the reaction: Glc1P ⟺ Glc6P), followed by the irreversible step mediated by ADPGlcPPase and then by reactions of starch synthase and branching enzyme (see also Iglesias and Podestá, 1996, for details). Also from intermediates of the RPPP, photoassimilates partition to the cytosol for sucrose synthesis via the Pi-triose-P counter-exchange (Iglesias *et al.*, 1993). From triose-P, fructose-6P is produced and then converted to glucose-6P by cytosolic PGI. The route is then continued to sucrose synthesis by a series of reactions (Iglesias and Podestá, 1996). Experiments carried out in *Clarkia xantiana* mutants, exhibiting decreased expression of cytosolic or chloroplastic PGI, showed different responses in carbon fluxes to starch and sucrose; thus demonstrating that the contribution of one enzyme is related to pathway in which it is involved, rather than with the reaction the enzyme catalyses. In these studies, it was found that a decrease of plastidic PGI inhibits (in high light but not in low light conditions) rates of sucrose build up as well as of photosynthesis. A decrease in cytosolic PGI instead reduces the rate of sucrose production (in low light but not in high light conditions), and this leads to a compensating increase of starch synthesis with no effect observed on the rate of photosynthesis (Stitt, 1999).

The analysis of the contribution of an enzyme to the flux of assimilates in a higher plant also needs to take into account a higher degree of compartmentation given by the existence of different cells (photosynthetic and heterotrophic). This was clearly exemplified in studies performed by Stark *et al.* (1992) on the genetic transformation of the metabolic pathway leading to starch synthesis in plant cells. It was shown that the increase in ADPGlcPPase activity within amyloplasts of potato (reached by the expression of a bacterial gene coding for a mutant, unregulated, ADPGlcPPase) produces plants that accumulate significant higher amounts of starch in their tubers (see below). In order to be successful in obtaining transgenic plants, that provide a product of higher quality (Collins and Shepherd, 1996), it was absolutely necessary to direct the transformant gene only to amyloplasts of the storage tissue (the gene was introduced under the control of a tuber-specific patatin; see Stark *et al.*, 1992). Otherwise, expression of the gene in a constitutive manner (i.e. expression under the control of CaMV 35S promoter) is detrimental for plant growth and development. The reason for this is that if ADPGlcPPase activity is increased in the chloroplast, an excess of starch is accumulated in the leaf, which then reduces sucrose synthesis and availability for export to actively growing tissues of the plant (Stark *et al.*, 1992). This demonstrates that ADPGlcPPase is a rate-controling step in starch synthesis provided is fully activated. Moreover, the different pathway followed by photoassimilates in source or sink tissues, implies a distinctive contribution of this enzyme to carbon flux and partitioning in the different plant cells.

Regulation and Control: Starch Synthesis, a Case Study

Starch is the major storage product of photosynthesis, as up to 30% of the carbon photoassimilated by plants is channeled to this polysaccharide under optimal conditions (Heldt *et al.*, 1977). Thus, knowing the regulation of starch levels in plants has a tremendous potential impact on the primary productivity on earth. We will refer to the manipulation of the starch biosynthetic pathway to produce a significant increase of its accumulation in crops.

Starch synthesis from glucose-1P in plants occurs through three reactions successively catalyzed by ADPGlcPPase, starch synthase (SS) and branching enzyme (BE) (Sivak and Preiss, 1998). In the first step, ADPGlc is produced from glucose-1P and ATP mediated by ADPGlcPPase. Then, different isoforms of SS utilize the sugar-nucleotide as the glucosyl donor to elongate an α-1,4-glucan chain; afterwards, specific branching isoenzymes introduce α-1,6-branch points in the polysaccharide molecule. The reaction catalyzed by ADPGlcPPase is irreversible *in vivo* and constitutes a key regulatory step in the pathway. Plant ADPGlcPPase is a highly regulated enzyme, the main allosteric effectors being 3PGA (activator) and Pi (inhibitor). The well known interplay between both regulatory metabolites of the enzyme has been recently highlighted after showing that ADPGlcPPase regulatory properties exhibit ultrasensitive behavior in molecularly crowded environments (Gómez-Casati *et al.*, 1999, 2000). Ultrasensitive systems exhibit sensitivity amplification (i.e. the percentage change in a response, compared with the percentage change in the stimulus). Amplification factors for ADPGlc synthetic rates ranging from 11- to 19-fold were described as a function of 3PGA levels. An advantage of this mechanism could be given by the fact that in the light-dark transition the system may be quickly deactivated (i.e. a sharp decrease of starch levels) thereby hindering wastage of ATP. In addition, starch synthesis during the dark-light transition would operate as an ultrasensitive switch-like device that can amplify 14- to 15-fold the levels of starch for a narrow 3PGA/Pi ratio.

Ultrasensitivity has been defined as the response of a system that is more sensitive to changes in the concentration of the ligand than is the normal hyperbolic response given by the Michaelis-Menten equation (Goldbeter and Koshland, 1981; Koshland *et al.*, 1982; Koshland, 1998). One main advantage of ultrasensitivity is, that as a device in biochemical networks, it allows a several-fold increase of the flux through a metabolic step over a narrow range of change in substrate (Koshland, 1998) or effector concentrations (Gomez-Casati *et al.*, 1999, 2000). In this sense, ultrasensitivity functions as a switch-like device that filters out small stimuli by restricting the range and threshold of stimulation (Ferrel, 1998).

Indeed, ADPGlcPPase is an enzyme that functions as an information-processing biochemical device (i.e. sensing the levels of its allosteric effectors: 3PGA and Pi) besides its well known role in synthesizing ADPGlc, the glucosyl donor for starch synthesis in plants, or glycogen in cynanobacteria. Thus, it becomes relevant to understand its dual information-mass transforming abilities quantitatively, along with the impact on polysaccharide synthesis regulation. Based on the *in vitro* under aqueous or polyethyleneglycol (PEG)-induced crowded environments (Gómez-Casati *et al.*, 1999, 2000), and *in situ* (Gomez-Casati *et al.*, 2001) cyanobacterial ADPGlcPPase kinetic and regulatory properties, a mathematical model was formulated that describes starch metabolism in cells performing oxygenic photosynthesis (Cortassa *et al.*, 2001, submitted; Aon *et al.*, 2001). Ultrasensitive behavior was also observed *in vitro* in ADPGlcPPase from spinach leaves (Gómez-Casati, Aon and Iglesias, unpublished results). The mathematical model takes into account the ADPGlc production by ADPGlcPPase as well as starch synthesis and degradation. Simulations performed with this model allowed to reproduce the ultrasensitive behavior of ADPGlcPPase and its corresponding amplification factors, under crowding conditions. Results from this model allow us to establish that polysaccharide synthesis is also ultrasensitive, a fact that had been previously suggested (Gómez Casati *et al.*, 1999) but not demonstrated (Gomez-Casati *et al.*, 2001, submitted; Cortassa *et al.*, 2001, submitted).

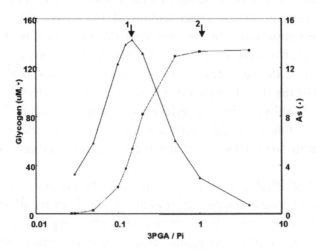

Figure 6.7. Steady state levels of storage polysaccharides and amplification factors exhibited by ADPGlucose pyrophosphorylase (ADPGlcPPase) as a function of the ratio 3PGA/Pi. Carbon metabolism from G1P towards storage polysaccharides (cyanobacterial glycogen or starch) synthesis and subsequent degradation, was modelled with a system of two ordinary differential equations (ODEs). The two equations represented the dynamics of ADPGlc and glycogen (GLY), i.e. the state variables, according to the following kinetic scheme:

$$ATP + G1P \xrightarrow{\ ADPGlcPPase\ } ADPGlc + PPi$$

$$ADPGlc + GLY_n \xrightarrow{\ GLYsy\ } GLY_{n+1} + ADP$$

$$GLY \xrightarrow{\ GLY\,deg\ }$$

V_{AGPase} represents the reaction rate catalyzed by *ADPGlcPPase*, and V_{GLYsy}, V_{GLYdeg} the glycogen synthesis and degradation, respectively. The ultrasensitive response of V_{AGPase} toward its allosteric activator 3 phosphoglyceric acid, 3PGA, elicited by orthophosphate, Pi, and polyethyleneglycol (PEG)-induced molecular crowding, operating under zero- or first-order conditions with respect to its substrates, G1P and ATP (Gomez- Casati *et al.*, 1999, 2000, 2001), was taken into account in the rate expression of the enzyme (Cortassa, Gomez-Casati, Iglesias and Aon, 2001, submitted).

The results of the model explained above, and shown in Fig. 6.7, clearly depict the links existing between regulation and control. There is a phase relationship between the maximum of amplification attained by ADPGlcPPase rate and the accumulation of starch, as a function of the 3PGA/Pi ratio. At arrow 1 (Fig. 6.7) the enzyme is maximally sensitive toward the stimulus given by the ratio between the two allosteric effectors. The maximal amplification implies a maximal relative increase in velocity in response to the stimulus. This is the Regulatory Phase; using the terminology of MCA the elasticity coefficients are maximal whereas the control coefficients are minimal (see Chapter 4). Arrow 2 points out a phase of behavior where the system is only minimally sensitive to the stimulus (Fig. 6.7). Otherwise stated, the enzyme is saturated thus becoming a rate-controling step. This is the Controling Phase.

A biochemical switch-like ADPGlcPPase can then function as a regulatory or control device depending upon the level of stimulus, entraining different polysaccharide levels. Indeed, at low 3PGA/Pi ratios increasing the cellular levels of ADPGlcPPase by the techniques of molecular biology will not translate into higher levels of starch, despite the potentiality of the enzyme to regulate the flux under those conditions. Higher 3PGA/Pi ratios should be necessary to allow the enzyme express its maximal potentiality. In good agreement with the latter, the successful transformation of plants to increase the amounts of starch accumulated in storage tissues, required the expression of a gene (*glgC 16*) coding for a mutant ADPGlcPPase from *E. coli*, insensitive to regulation and exhibiting high specific activity (Stark *et al.*, 1992; Collins and Sheperd, 1996). Thus in transgenic plants, the pathway of starch synthesis operates under non-regulated conditions and continuously in the phase of behavior minimally sensitive to the stimulus, with ADPGlcPPase activity maximized and steadily controling the flux (Fig. 6.7, arrow 2). The result was that transgenic plants accumulated over 30% higher amounts of starch. Interestingly, the expression of an ADPGlcPPase sensitive to regulation showed no modification in the levels of polysaccharide (Stark *et al.*, 1992). In other words, enhancing the amount of enzyme in the Controling Phase was effective for increasing the amount of final product, as expected according to the analysis presented in Fig. 6.7. All these results demonstrate that, when fully activated, ADPGlcPPase may be a main rate-controling step of starch synthesis.

Concluding Remarks

MCE as applied to plants has been hardly developed, although at present receives considerable interest. Photosynthetic organisms (especially higher plants) are complex biological systems mainly due to the occurrence of multiple metabolic routes (pathway duplication) with each path operating in different intracellular compartments. In fact, although much relevant information on plant metabolism (and its regulation) has been obtained, the complete understanding of the complex scenario is far from complete.

Plants constitute highly appropriate organisms to be employed for MCE; because of their autotrophy they potentially constitute self-sufficient biological factories. Techniques for their *in vitro* growth as well as for the genetic transformation have been developed, and many important biotechnological goals have utilized plants for obtaining novel or higher quality products. It is envisaged that special efforts will focus on further characterization of fluxes of matter and energy and metabolic control and regulation in plant cells. This is a very necessary step for optimising conditions and tools for rationally improving the performance of plants within different biotechnological processes. The whole scenario looks exciting, both for the possibility of accomplishing a better understanding of plant metabolism and physiology, as well as the many relevant applications deriving from such knowledge.

Chapter 7

Cellular Engineering

Outline

In the last thirty years of research on metabolism, a great deal of experience and understanding has been obtained by a combination of the theoretical and experimental developments as described all throughout this book.

Of recent refinements to the experimental approach, the combination of fermentation and DNA recombinant technologies is perhaps the most powerful (Weusthuis *et al.*, 1994; Vriezen and van Dijken, 1998; Aon and Cortassa, 1997, 1998; Cortassa and Aon, 1997, 1998). Metabolic Control Analysis (Kacser and Burns, 1973; Heinrich and Rapoport, 1974; Kell and Westerhoff, 1986; Fell, 1992), Metabolic Flux Analysis (Vallino and Stephanopoulos, 1993; Varma and Palsson, 1994; Zupke and Stephanopoulos, 1995; Cortassa *et al.*, 1995; Jorgensen *et al.*, 1995; Vanrolleghem *et al.*, 1996; Bonarius *et al.*, 1996) and Biochemical Systems Theory (Savageau, 1976, 1991; Cornish-Bowden, 1989) along with the classical studies emerging from the so called "black-box approach" (Roels, 1983), microbial biochemistry and physiology (Stouthamer, 1979; Ingraham *et al.*, 1983) and Mosaic Non Equilibrium Thermodynamics (Westerhoff and van Dam, 1987) are the major theoretical achievements which have provided mathematical modeling and conceptual tools. Another powerful tool for analyzing the dynamic behavior of cells, biochemical pathways or biological systems in general, is the Dynamic Bifurcation Theory (DBT) (Doedle, 1986; Abraham, 1987). A Biothermokinetic Method based on a numerical combination of DBT and MCA has been proposed (Aon and Cortassa, 1991; Cortassa *et al.*, 1991; Aon and Cortassa, 1997). This approach allows us to quantitate the behavior of metabolic pathways in both kinetic thermodynamic terms and also to predict new types of behavior.

Although there is general agreement about the essential unity of life from a biochemical and genetic points of view, there is not yet consensus on whether that unity extends to the global rules of functioning of metabolic networks. This problem concerns the function of cells. There is compelling evidence in the context of the common stages in the life of a cell (e.g. growth, proliferation and differentiation) that these putative general rules do exist and may be applied to either prokaryotic or eukaryotic cells.

The Global Functioning of Metabolic Networks

We have emphasized that even in their integrated form, the genomic databases will not allow us to go from DNA sequence to function in a straightforward way. This is for two main reasons: *(i)* the ability of cells to become dynamically organized under homeodynamic conditions; *(ii)* the dependence of phenotype on the spatio-temporal display of genetic information in mass-energy-information-carrying networks of reactions that provide the essence of organized complexity (see Chapter 1).

Biological organized complexity is characterized by its simultaneous functionality at several levels of organization spanning a broad range of spatio-temporal scales, and its ability to generate emergent phenomena, realized as spatio-temporal structures (Lloyd, 1992; Aon and Cortassa, 1997). These are common features exhibited by different cellular systems, and their basic dynamic nature, allow us to conclude that not all the information concerning the spatio-temporal organization in cells and organisms is encoded in the DNA. Furthermore, the putative rules that govern the global behavior of the mass-energy-information carrying networks of cells, result from the emergent properties of these networks themselves. Cells exhibit many distinct signaling (information carrying) and metabolic (mass and energy conversion) pathways strongly associated with the dynamic cellular scaffolds (Aon *et al.*, 2000) that allow them to react to environmental stimuli (Fig. 1.6). This networking implies that the arrangement of a large number of components comprising their interactions as well as spatial relationships are critical (Weng *et al.*, 1999); function results in several emerging properties that the individual pathways do not themselves have. In this way information is stored within intracellular biochemical reactions of signaling pathways through multiple interactions that arise from protein-protein association, feedbacks (such as substrate inhibition), feedforwards (product activation), cross-activation or cross-inhibition. Moreover, chemical reactions organize themselves into linear, branched, and cyclic arrays, or as combinations of these basic topologies, and this in turn results in specific patterns of flux control, energetic performance, dynamic behavior, and signaling features (see Chapter 1, Fig. 1.7).

In the following, we describe some examples in which global functioning of cells and their metabolic networks is manifest in real biological contexts.

The Nature of the Carbon Source Determines the Activation of Whole Blocks of Metabolic Pathways with Global Impact on Cellular Energetics

The fermentative (e.g. glucose, fructose) or non-fermentative (gluconeogenic, e.g. glycerol, acetate, ethanol) nature of the carbon source determines the choice of catabolic pathways. For gluconeogenic substrates, catabolism relies largely on the functioning of the TCA cycle and respiratory chain, whereas growth on glucose requires catabolism through fermentative pathways (Table 4.1). A consequence for cellular energetics is that the ATP requirement for the synthesis of cell material, is profoundly influenced by the nature of the C and N source. A similar effect is exhibited by the C assimilatory pathway (Stouthamer and van Verseveld, 1985).

Moreover, in yeast, the well-known pleiotropic phenomenon of glucose-induced catabolite repression implies the inactivation of the expression of whole sets of genes (Fig. 4.8). The products of these genes correspond to enzymes related either to the degradation of other carbon sources, or with proteins involved in the regulation of derepressing mechanisms in the presence of gluconeogenic carbon sources.

The redirection of metabolic pathways away from carbohydrate to aromatic amino acid synthesis activates a new cycle that involves the enzyme phosphoenol pyruvate carboxykinase (Pck). The latter implies that the glyoxylate cycle may replace the TCA cycle with only a small decrease in efficiency (Liao *et al.*, 1996). Gene overexpression corresponding to enzymes that synthesize PEP (Pck, phosphoenol pyruvate synthase, Pps) stimulates the rate of glucose consumption, represses heat shock, and negatively regulates the nitrogen assimilation regulon (Ntr regulon). It has been suggested that these three effects could be mediated by a glycolytic intermediate (Liao *et al.*, 1996).

Carbon Sources that Share Most Enzymes Required to Transform the Substrates into Key Intermediary Metabolites under Similar Growth Rates, Bring About Similar Fluxes through the Main Amphibolic Pathways

Fluxes through the central metabolic pathways of *S. cerevisiae* have been quantitated by MFA (Cortassa *et al.*, 1995; Aon and Cortassa, 1997) under balanced growth conditions in the presence of fermentative (glucose) or gluconeogenic (acetate, pyruvate, lactate, ethanol, glycerol) substrates. Similar fluxes through the central pathways were exhibited by yeast in the presence of pyruvate and lactate or ethanol and acetate that share most metabolic pathways for their assimilation and also result in similar growth rates (Table 4.1)

As expected, widely different fluxes through specific enzymatic steps were determined when pathways for utilization of the carbon source differ. This was the case of, e.g. acetate, lactate and glucose. A large flux through the AcCoA synthase-catalyzed step was required by acetate as well as an extensive use of the glyoxylate cycle and of the NADP-dependent malic enzyme to supply key intermediary metabolites. On the other hand, lactate sustained large fluxes through the pyruvate carboxylase step, which was also used to a lesser extent by glucose or glycerol to replenish OAA and αKG, although not required at all for growth on acetate (Cortassa *et al.*, 1995).

Interaction between Carbon and Nitrogen Regulatory Pathways in S. cerevisiae

Gluconeogenesis from glutamate as N-source, may simultaneously operate with a high glycolytic flux in glucose-limited chemostat cultures (Aon J.C. and Cortassa, 2001). During growth on fermentable carbon sources such as glucose, most gluconeogenic enzymes, e.g. phosphoenolpyruvate carboxykinase (PEPCK), fructose 1,6 bisphosphatase and malic enzyme, are repressed at the transcriptional level. Nevertheless, the latter has always been assayed in media with ammonia as N-source, or in rich nutrient agar with glutamine. In the presence of glutamate, we found that *S. cerevisiae* displayed an almost complete fermentation of the glucose consumed with biomass being synthesized from glutamate under C-limiting conditions in the respiro-fermentative growth mode of glucose-limited chemostat cultures (Figs. 7.1 and 3.9). The latter implies that the enzymes of the gluconeogenic pathway, namely PEPCK and the malic enzyme are not repressed (activity levels of 22 and 49 nmol min^{-1} mg^{-1} protein, respectively, were measured). This situation contradicts the traditional concept of catabolite repression, further suggesting the interplay between carbon and nitrogen regulatory pathways. The derepression of gluconeogenic enzymes could be perhaps elicited by the presence of an N-source such as glutamate. This amino acid also provoked a much higher qO$_2$ under N-limitation than that displayed in ammonia-fed cultures (Aon, J.C. and Cortassa, 2001).

At the molecular level, a dual regulation of NADP-GDH by C and N sources has been reported, suggesting an interaction between both routes of metabolism, although the physiological consequences of this dual control were not addressed (Coshigano *et al.*, 1991). The simultaneous operation of glycolysis and gluconeogenesis suggests that the dual regulation of a nitrogen repressible gene such as *GDH2* may be extended to gene products expressed differentially according to carbon catabolite repression. A dual regulation at biochemical and genetic levels was proposed for glutamine synthesis that would act as a regulatory signal of glucose breakdown through the ATP/ADP ratio (Flores-Samaniego *et al.*, 1993). The box-binding HAP transcription factor has also been proposed as a new regulatory crossroad between nitrogen and carbon metabolism (Dang *et al.*, 1996). From the results of Aon, J.C. and Cortassa (2001), we hypothesize that both nitrogen and carbon catabolite repression are interlinked phenomena and that many genes under the control of each of these general mechanisms are subjected to a dual control by C and N sources.

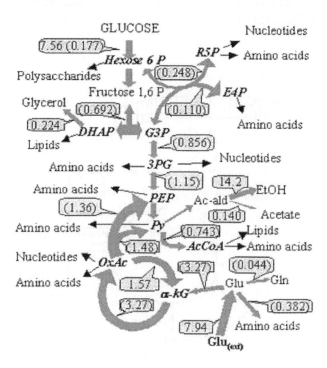

Figure 7.1. Metabolic fluxes determined or estimated in *S. cerevisiae* in the presence of glutamate as N source, during the respiro-fermentative regime of C-limited aerobic chemostat cultures. The flux values at *D*=0.36 h⁻¹ are shown. The fluxes experimentally determined are indicated without parenthesis in the oval, whereas estimated ones are depicted between brackets. The ovals point to the metabolic step for which the flux has been estimated or measured. The thickness of the arrows attempt to give a qualitative indication of the magnitude of the overall flux (anabolic plus catabolic) through a given reaction step, or lumped reaction steps within a metabolic pathway. Monomer precursors of macromolecules such as aminoacids and nucleotides, polysaccharides, or lipids, as final products of the anabolism are emphasized on a dashed background. (Reproduced from *Metabolic Engineering*, Aon, J.C. and Cortassa, ©copyright 2001 Academic Press, with permission).

Flux   Redirection   toward   Catabolic   (Fermentation)   or   Anabolic (Carbohydrates) Products May Be Generated as a Result of Alteration in Redox and Phosphorylation Potentials

During transitions from aerobic to oxygen-limited or anaerobic conditions, the correct   balance   of   both   redox   (NADH/NAD;   NADPH/NADP)   and phosphorylation (ATP/ADP) couples is of great importance to maintain functional metabolic pathways: we cite two examples.

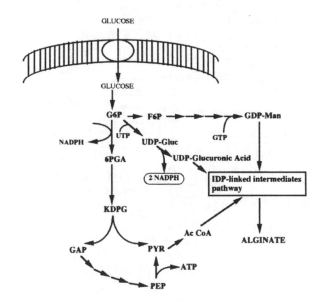

Figure 7.2. Pathways of sugar breakdown (Entner-Doudoroff) and alginate synthesis in *Pseudomonas mendocina*. The production of the activated sugars (UDP-Glucuronic acid) leading to alginate synthesis, renders additional reduced coenzymes (NADPH) that are oxidized in the electron transport chain (Adapted from Hunt and Phibbs, 1983, and Jarman and Pace, 1984). Under microaerophillic conditions, *P. mendocina* redirects more than 50% of the carbon consumed toward alginate production (Verdoni *et al.*, 1992). NADPH production and its oxidation in a higher efficient electron transport chain (P/O=3) allow to keep the redox as well as phosphorylation potentials, to a level high enough to proliferate at 0.25 h^{-1} in nitrogen-flushed chemostat cultures.

Indeed,   the   absence   of   electron   acceptors   introduces   alterations   in phosphorylation potentials as well as redox potentials. In continuous culture, under microaerophilic conditions, *Pseudomonas mendocina* growing in synthetic medium exhibited fermentative metabolism with excretion of ethanol as well as lactic, formic and acetic acids. The amounts of fermentation product detected in the extracellular medium suggested that the organism attained its redox balance by lactic acid and ethanol production (Verdoni *et al.*, 1992; Aon and Cortassa, 1997). Under those microaerophilic conditions, this organism could redirect, not only catabolic pathways, but also induce new anabolic pathways such as those of uronic acids or alginate synthesis (Fig. 7.2). In fact, around 50% of the carbon was found in the extracellular medium as uronic acids (Verdoni *et al.*, 1992).

Intracellular accumulation of polyhydroxybutyrate (up to 50% of its dry weight) occurs under oxygen limitation in *Azotobacter beijerincki* (an aerobe that catabolizes glucose via the Entner-Doudoroff pathway like *Pseudomonas*) (Senior *et al.*, 1972; Anderson and Dawes, 1990).

Temperature-Dependent Expression of Certain Mutations Depend upon the Carbon Source

Gluconeogenic substrates induced suppression of the terminal phenotype of the cell division cycle mutant *cdc*28 at the restrictive temperature. *cdc*28 was chosen because its terminal phenotype shows a clear and aberrant change in shape called "shmoo" (Hartwell, 1974). When the shift to the restrictive temperature was performed in the presence of 1% acetate or 1% glycerol, the "shmoo" phenotype did not appear for either of these gluconeogenic substrates. The terminal phenotype suppression by glycerol or acetate was reversible, since *cdc*28 yeast cells grown either in glycerol or acetate when readapted to glucose, did show the "shmoo" at the restrictive temperature (Aon and Cortassa, 1995, 1997). This result clearly shows an environment-induced change in cell shape. By directly acting on metabolic pathways through challenging the cell to utilize different pathways from those associated with cell proliferation arrest, i.e. fermentative, we were able to block the change in cell shape (shmoo).

There Seems to Exist a General Pattern of Control of the Intracellular Concentration of Metabolites

By applying MCA it has been deduced that metabolite concentrations are controlled by the rate-controling steps of the flux (Cortassa and Aon, 1994a). In addition, metabolite levels are negatively controlled by the enzyme consuming that metabolite (Table 7.1) (see also Chapter 4).

The positive control exerted upon the level of a metabolite M1 by a certain step of the pathway, means that an increase in substrate consumption results in a rise of M1 at the steady state. On the contrary, a negative control coefficient entrains a decrease of M1 at the steady state. For example, as shown in Table 7.1, the rate-controling steps of the flux at $D = 0.225$ h^{-1} are: sugar uptake (c_{IN}^{PEP} =0.12), the HK (c_{HK}^{PEP} =0.701) and PFK (c_{PFK}^{PEP} =0.231) that exert a positive control on the PEP concentration, and the ATPase with a negative control on the PEP concentration ($C_{ATPase}^{PEP} = -1.35$).

Table 7.1. Metabolite concentration control coefficients of phosphoenolpyruvate (PEP) and ATP by the different glycolytic steps, in the wild type strain CEN.PK122 and *snf*1 mutant growing in aerobic, glucose-limited chemostat cultures. The control coefficients were obtained from the **E** matrix inversion, as described in the text (see Chapter 4). The model of glycolysis and the branch toward the TCA cycle and ethanolic fermentation, are depicted in Fig. 4.10.

Strain $D(h^{-1})$	C_{IN}^{PEP}	C_{HK}^{PEP}	C_{PFK}^{PEP}	C_{ALD}^{PEP}	C_{GAPD}^{PEP}	C_{PGK}^{PEP}	C_{PK}^{PEP}	C_{ADH}^{PEP}	C_{TCA}^{PEP}	C_{ATPase}^{PEP}
WT										
0.1	0.132 (0.06)	0.772 (0.35)	1.34 (0.608)	-0.195	-0.008	0	-1.33	-0.415	0.142	-0.433
0.225	0.442 (0.12)	2.60 (0.701)	0.86 (0.231)	-0.115	-0.049	0	-1.54	-1.30	0.447	-1.35
0.3	0.012 (0.008)	0.012 (0.008)	1.51 (0.98)	-0.029	-0.008	0	-1.58	-0.303	1.38	-1.00
*Snf*1										
0.05	0.007 (0.003)	0.011 (0.006)	2.02 (0.003)	-0.046	-0.018	0	-1.45	-0.277	0.095	-0.335
0.15	0.254 (0.13)	0.803 (0.41)	0.96 (0.49)	-0.034	-0.007	0	-1.60	-0.356	0.425	-0.445
0.25	-0.089 (0.021)	-0.081 (0.02)	-3.79 (0.996)	-0.013	-0.014	0	-2.04	-0.305	7.45	-1.10

Strain $D(h^{-1})$	C_{IN}^{ATP}	C_{HK}^{ATP}	C_{PFK}^{ATP}	C_{ALD}^{ATP}	C_{GAPD}^{ATP}	C_{PGK}^{ATP}	C_{PK}^{ATP}	C_{ADH}^{ATP}	C_{TCA}^{ATP}	C_{ATPase}^{ATP}
WT										
0.1	0.097	0.571	0.991	0	0	0	0	-0.975	0.335	-1.02
0.225	0.194	1.14	0.378	0	0	0	0	-1.01	0.35	-1.05
0.3	-0.0006	-0.0006	-0.079	0	0	0	0	-0.303	1.38	-1.00
*Snf*1										
0.05	0.005	0.009	1.55	0	0	0	0	-0.837	0.287	-1.01
0.15	0.11	0.35	0.417	0	0	0	0	-0.83	0.99	-1.04
0.25	-0.12	-0.107	-4.99	0	0	0	0	-0.264	6.44	-0.96

Another distinctive feature shown in Table 7.1 is that metabolites concentration with a conversion step involving other metabolites, is additionally controlled by the enzymes catalyzing the conversion steps of those other intervening metabolites. This is clearly the case for PEP, whose concentration is controlled by aldolase (negatively), alcohol dehydrogenase (negatively), ATPase (negatively) and by the functioning of the TCA cycle, positively.

Table 7.2. Intracellular metabolite concentrations determined in glucose-limited chemostat cultures of *S. cerevisiae*. Metabolite concentrations are expressed in mmol l^{-1} of intracellular volume.

Strain $D (h^{-1})$	ATP	ADP	AMP	NAD	G6P + F6P	FdP	DHAP + G3P	DPG	PEP	Pyr
WT										
0.1	3.05	1.249	1.379	0.3650	8.36	0.44	0.44	2.50	2.21	< 0.1
0.225	2.00	0.608	0.072	0.0276	4.06	0.51	0.73	3.35	2.87	< 0.1
0.3	2.71	0.846	0.174	0.0046	35.65	1.43	1.06	4.50	3.64	5.98
*snf*1										
0.05	0.79	0.719	0.440	0.0092	60.56	0.94	1.15	3.32	2.66	< 0.1
0.15	0.88	0.802	0.883	0.1850	5.06	1.20	1.30	3.76	3.64	1.62
0.25	0.95	0.538	1.200	0.0306	19.11	2.41	2.04	7.57	5.47	31.40
snf4										
0.1	2.46	0.733	0.347	0.3328	1.90	1.03	1.41	2.95	2.96	0.53
0.2	4.37	1.319	0.556	0.5543	1.21	1.14	1.39	4.22	3.93	0.33
0.3	4.68	1.444	0.320	0.6923	0.24	3.18	1.77	6.80	5.72	24.81
*mig*1										
0.1	2.72	0.940	0.339	0.3039	2.20	1.81	1.02	2.98	2.61	< 0.1
0.2	3.74	1.589	0.424	0.4848	0.89	5.41	2.15	5.93	5.10	0.29
0.3	4.32	2.506	0.675	0.2577	1.78	4.79	1.88	6.46	5.00	8.16

Table 7.3. Main metabolic fluxes sustained by the wild type strain and catabolite-repression mutants in glucose-limited chemostat cultures of *S. cerevisiae*. The fluxes are expressed in mmol h^{-1} g^{-1} dw.

Strain		qO_2	qCO_2	qGlc	QetOH
	D (h^{-1})				
WT					
	0.1	2.12	2.38	1.22	0.0442
	0.225	4.31	6.69	3.87	1.7531
	0.3	4.85	11.15	7.51	5.7928
*snf*1					
	0.05	2.11	2.74	1.04	0.0097
	0.15	4.03	5.71	3.78	0.3558
	0.25	6.34	11.55	7.23	1.1592
*snf*4					
	0.1	2.54	2.44	1.37	0.0142
	0.2	5.91	10.41	4.13	1.7488
	0.3	5.76	16.58	6.99	4.9086
*mig*1					
	0.1	3.16	3.55	1.48	0.0113
	0.2	3.88	5.58	4.21	1.8335
	0.3	3.93	13.41	10.72	7.1256

Reprinted from *Enzyme and Microbial Technology*, 21, Cortassa and Aon, Distributed control of the glycolytic flux in wild-type cells and catabolite (de)repression mutants of *Saccharomyces cerevisiae* growing in carbon-limited chemostat cultures, 596-602. ©copyright 1997, with permission from Elsevier Science.

The *snf*1 mutant when grown in glucose-limited chemostat cultures exhibited low biomass yields and specific rates of ethanol production, along with high rates of respiration (Cortassa and Aon, 1998) (Table 7.3). The three- to four-fold lower ATP and ADP concentrations and the ability to accumulate large amounts of intracellular G6P with respect to the wild type strain, indicated altered energetic metabolism of the *snf1* mutant (Cortassa and Aon, 1997; Cortassa and Aon, 1998; Aon and Cortassa, 1998) (Table 7.2). This mutant shows strikingly different patterns of control coefficients in two tightly controlled metabolites such as PEP and ATP (Table 7.1) especially at high growth rates. It is remarkable the increase and change of sign (from positive to negative) of the PFK concentration control coefficient on either PEP or ATP. This change may be rationalized as follows: an increase in enzyme activity (PFK) will effect a decrease in a metabolite concentration (PEP or ATP) (a negative control coefficient) leading in turn to higher FDP levels (a PK allosteric activator) and in that way decreasing PEP levels.

Equally remarkable was the increase of the absolute values of the TCA cycle control coefficients on both metabolites. The strong positive effect on the control coefficient of the ATP concentration (almost 7-fold) is due to an increase in the functioning of the TCA cycle that in turn gives higher levels of ATP. The increase of the TCA cycle and the ATP levels decreases ADP ones along with PK activity that allows the increase of PEP levels (almost 20-fold). The drastic increase of the control exerted by the TCA cycle on PEP concentration in the *snf1* mutant is due to the basic autocatalytic nature of the glycolytic pathway and, particularly, to the cooperative nature of PK activity. It must be remembered that the fermentative mode of glycolysis is strongly active at high growth rates in *S. cerevisiae*.

Dependence of the Control of Glycolysis on the Genetic Background and the Physiological Status of Yeast in Chemostat Cultures

It has been shown that the rate-controling steps of glycolysis may be redistributed between the glucose transport, PFK, and HK, depending upon the catabolite repression properties of *S. cerevisiae*, and the regime of glucose breakdown, in aerobic, glucose-limited chemostat cultures (Fig. 4.11). The control analysis was performed at different physiological regimes, i.e. oxidative, fermentative, or just after the onset of ethanolic fermentation.

A series of isogenic mutants of catabolite repression on the background of the wild type strain CEN.PK122 were utilized. These mutants carry disruption in genes involved in the expression of glucose-repressible genes, namely *SNF1/CAT1/CCR1*, *SNF3/CAT3*, and *MIG1*. Their metabolic and physiological behavior was compared with that of the wild type strain (Cortassa and Aon, 1997, 1998) (Tables 7.2 and 7.3; see also Chapter 4). The lack of proteins Snf1, Snf4, and Mig1 involved in glucose repression or release of repression in gluconeogenic carbon sources (Fig. 4.8), elicit a particular metabolic state that is reflected by the physiological behavior of each mutant (Fig. 4.9) and the pattern of control coefficients (Fig. 4.11). The onset of ethanolic fermentation was different depending on the mutant. In fact, the critical dilution rates, Dc, at which ethanol production was triggered were: 0.2, 0.1, 0.15, and 0.17 h^{-1}, for the wild type, snf1, snf4, and the mig1 mutant, respectively. Thus, the catabolite repression features of yeast regulate the overall metabolism, which in turn translates into the control of glycolysis (Cortassa and Aon, 1997, 1998; Aon and Cortassa, 1998). The relative weight of the control exerted by the main rate-controling steps changed according to the catabolite (de)repression properties of the mutants (Fig. 4.11), and their physiological (Table 7.3) as well as metabolic (Table 7.2) status of yeast (Cortassa and Aon, 1997, 1998).

Remarkably, the snf1 mutant showed pleiotropic effects in its pattern of metabolic (Table 7.2) and physiological (Table 7.3) behavior. A low fermentative ability along with low intracellular levels of adenine nucleotides, and high concentrations of G6P, allow to explain the control exerted on glycolysis flux by the branch to the TCA cycle (Figs. 4.10 and 4.11). Likely, the insufficient ethanolic fermentation makes snf1 mutant more dependent on the ATP provided by respiration. This is in agreement with the unusual ability of snf1 to increase its respiratory flux even above Dc (Table 7.3) (Cortassa and Aon, 1997).

Cellular Engineering

In previous sections (see: *The flux coordination hypothesis*, Chapter 3, and *The global functioning of metabolic networks*, this chapter), it was argued that cells exhibit global regulatory mechanisms of functioning. In the framework of a dynamic view of cell function, growth along with cell division and differentiation, have been considered as a coordinated, dissipative balance of fluxes. The latter results from fluxes of the different materials required, the interactions between those fluxes, and the properties emerging from such interactions.

Particularly, the Flux Coordination Hypothesis emphasizes the regulation of the degree of coupling between catabolic and anabolic fluxes as a global regulatory mechanism of the growth rate, and the rate of energy dissipation (Figs. 3.4 and 3.5) (Aon and Cortassa, 1995, 1997). Our aim here is to stress that we may take advantage of these global regulatory mechanisms exhibited by cells for engineering them. Directly influencing these global mechanisms of cell functioning, is what we call Cellular Engineering. Attempts in this direction have been already reported in the literature as applied to mammalian cells (CHO, Chinese hamster ovary) through one-step multigene engineering (Fussenegger *et al.*, 1998, 1999). In CHO cells the main targets of engineering have been the control of: *(i)* the operation of the cell cycle in cultured cells; and *(ii)* cellular apoptosis.

Growth Rate, G1 Phase of the Cell Cycle, Production of Metabolites and Macromolecules as Targets for Cellular Engineering

Cellular bioenergetics and the growth rate of microorganisms.

There is a fundamental relationship between the growth rate of yeast and the duration of quiescence at the G1 stage of the mitotic cycle (Fig. 3.6), phase most sensitive to growth conditions (reviewed in Aon and Cortassa, 1995). Thus cell growth (i.e. macromolecules synthesis) and proliferation (i.e. cell division) are somewhat coupled.

A general coupling mechanism has been proposed in the framework of FCH, according to which a certain imbalance between anabolic and catabolic fluxes will affect the fate of a cell (Fig. 3.5). When a cell is challenged with different environmental stimuli (substrate shortage, inhibitors, pH or temperature changes), or changes in its genetic make up (e.g. by gene overexpression or expression of heterologous proteins), or undergoes a commitment to divide or to follow a certain developmental path, the general pattern of metabolic fluxes changes (Aon and Cortassa, 1995; 1997; Aon, J.C., 1996; Aon, J.C. and Cortassa, 1996; Aon J.C. *et al.*, 1996, 1997). The new emerging pattern of fluxes is accompanied by a coordinate increase or decrease of the amounts of all the enzymes involved (Srere, 1993; Aon, J.C., 1996; Aon, J.C. *et al.*, 1996; Aon and Cortassa, 1997; Cortassa *et al.*, 2000).

The expression of genes that code for enzymes of gluconeogenic, glyoxylate and oxidative phosphorylation pathways in yeast, is derepressed in the presence of gluconeogenic carbon sources and in the absence of glucose (Celenza and Carlson, 1989; Schuller and Entian, 1991). On the contrary, glucose represses the expression of those enzymes and favors glucose breakdown through the glycolytic pathway. In the realm of mammalian biochemistry, when rats are starved and re-fed a high carbohydrate diet, all the enzymes of fatty acid synthesis are increased. Injection of 3,5,3'-triiodethyronin (T3) (or prolonged exercise), increases all the amounts of enzymes of the TCA cycle in rat muscle as well as proteins of electron transport (Srere, 1993, and ref. therein).

The FCH states that flux (im)balance is reflected by the general carbon and energetic coupling between catabolic and anabolic fluxes, which in turn mediate subcellular remodeling and cellular transformation. A way to produce flux imbalance or alteration of coordination between catabolic and anabolic fluxes is to over- or under-produce a given protein or a metabolite (e.g. by expression of specific foreign proteins or by defective mutants, respectively) (see Aon and Cortassa, 1995, 1997, for reviews). Temperature-sensitive cell cycle mutants of *S. cerevisiae* bearing mutations defective in cdc gene products, have either catabolic or anabolic fluxes and their degree of coupling impaired concomitantly with the arrest of cell proliferation (Aon *et al.*, 1995; Aon and Cortassa, 1995, 1997).

Cellular flux imbalances introduced by foreign gene expression should have consequences at the level of metabolic coupling, which is somehow reflected at the cellular level. Prokaryotic and eukaryotic cells are known to react to unbalanced overproduction of protein (mediated either by vectors, malignant transformation or by drugs) with changes in their morphology. If the unbalanced flux is not directed to biomass synthesis, it results in morphological alterations with respect to a standard cell shape. Several reported data show that these are indeed the case and have been extensively reviewed elsewhere (Aon and Cortassa, 1995, 1997).

Detection and quantification of the contribution exerted by large metabolic blocks, and pathways thereof, to global cellular flux imbalance.

During arrest of cell proliferation, flux redirection was found to be associated with carbon and energetic uncoupling (Aon *et al.*, 1995; Mónaco *et al.*, 1995). In the framework of the FCH, when cells are challenged by an unfavorable environmental condition (e.g. temperature), distinct metabolic and energetic requirements are induced leading to differential gene expression, metabolic flux redirection, that in turn induce lower growth rates (Aon and Cortassa, 1997). Indeed, cell arrest of proliferation at 37°C (restrictive temperature) in the cell division cycle (*cdc*) mutants of *S. cerevisiae* *cdc*28, *cdc*35, *cdc*19, *cdc*21, *and* *cdc*17, was correlated with carbon and energy uncoupling (Aon *et al.*, 1995; Aon and Cortassa, 1995, 1997; Mónaco, 1996). Apparently, the flux imbalance associated with cell division arrest was mainly due to unregulated catabolic fluxes being more dramatic for carbon than for energy. At 37°C, cdc mutants diverted to biomass synthesis only 3% and around 20% of the fluxes of carbon consumed and ATP obtained by catabolism, respectively, compared with 50% and 30% in the wild type strain A364A. At 37°C, *cdc* mutants directed 60% of the carbon to ethanol production. *cdc*28 and *cdc*35 showed decreased mitochondrial biogenesis as well as impaired respiration (Genta *et al.*, 1995). When the anabolic fluxes sustained by *cdc*28, *cdc*35, *and cdc*21 were analyzed through incorporation of radioactive precursors into macromolecules (DNA, RNA, proteins), similar specific rates of incorporation were obtained in *cdc* mutants with respect to the wild type at the restrictive temperature (Table 7.4). (Aon and Mónaco, 1996; Mónaco, 1996). The chemical composition of *cdc*35 and *cdc*21 mutants at either permissive (25°C) or restrictive temperatures did not differ significantly with respect to the wild type (Table 7.5) (Aon and Mónaco, 1996; Mónaco, 1996; Mónaco and Aon, unpublished results). When analyzed together, the results obtained are in agreement with a major contribution of catabolism to the carbon and energetic imbalance associated with the arrest of cell proliferation in *cdc* mutants.

Within the large block of catabolism, the ethanolic fermentation pathway was shown to be activated and associated with the carbon and energetic uncoupling exhibited by the *cdc* mutants during cell proliferation arrest (Aon *et al.*, 1995). Further experimental evidence suggested that the increased fermentative ability, and the drastic carbon uncoupling, must be somehow linked to glucose-induced catabolite repression (Mónaco *et al.*, 1995). This is the topic of the next section.

Catabolite Repression and Cell Cycle Regulation in Yeast

The eukaryotic cell cycle comprises two main stages, mitosis (M) and DNA synthesis (S), separated by two gap periods, G1 and G2 (Lloyd *et al.*, 1982; Lloyd, 1998). Genetic and molecular biological approaches have contributed a great deal to the understanding of cell cycle regulation (Hartwell, 1991; Surana *et al.*, 1991; Reed, 1992). Particularly important has been the discovery that the CDC28 gene product of *S. cerevisiae*, the CDC2 gene product of *Schizosaccharomyces pombe*, and the maturating promoting activity of *Xenopus* were all related serine-threonine protein kinases (Hartwell, 1991). Furthermore, it has also been shown that the kinases of *S. pombe*, *S. cerevisiae*, and humans are functionally homologous after transfer genetic experiments (Forsburg and Nurse, 1991). Therefore, all the evidence available suggests that the mechanisms of cell cycle regulation are likely to be shared even in distantly related species.

Table 7.4. Average specific incorporation of radioactive precursors into macromolecules during eight hours at the restrictive temperature in wild type *S. cerevisiae* cells and mutants *cdc*21 and *cdc*35.

Strain	Specific incorporation into macromolecules (cpm mg^{-1} dw h^{-1})		
	Proteins	RNA	DNA
WT	6592	1613	908
*cdc*21	7883	1997	1001
*cdc*35	5657	1551	719

Reproduced from Mónaco, 1996, PhD thesis, Universidad Nacional de Tucuman, Argentina.

As shown in Fig. 3.6, further increases in growth rate trigger an exponential shortening of the G1 phase in yeast grown under batch (Fig. 3.6 A) or chemostat

(Fig. 3.6 B) culture conditions. Thus, there is a linear relationship between the temporal length of G1 and the doubling time. Interestingly, the specific rate of production of homologous proteins also depends on the growth rate (see next section). Under the same conditions, the combined length of the cell cycle periods S+G2+M only slightly declines with the growth rate, especially at low growth rates. The sort of limitation to which the cells are subjected under continuous culture appears also to influence the cells' distribution in the cell cycle. For instance, the response of the cells to nitrogen limitation in chemostat cultures, was to stay longer in G1 (Aon and Cortassa, 1995, 1997).

Table 7.5. Chemical composition of strains WTA364A, *cdc*21 and *cdc*35 at the restrictive temperature in 1% glucose and minimal medium.

Strain		RNA	Carbohydrates	Proteins	Lipids
		(%)	(%)	(%)	(%)
WT					
	t0	13.08±0.99	22.2±1.8	43.65±1.9	17.07±4.6
	t4	10.70±2.1	30.55±1.2	42.1±3.1	13.65±6.4
	t8	8.73±1.9	25.85±2.9	40.75±1.7	15.8±6.5
*cdc*21					
	t0	13.05±1.6	29.95±1.7	47.25±1.7	10.1±5
	t4	9.37±1.7	29.5±0.8	47.75±1.1	11.37±3.6
	t8	7.8±0.95	27.25±3.2	45.2±1.7	16.25±5.85
*cdc*35					
	t0	12.15±2.0	35.7 ±1.1	34.9±2.3	13.25±5.4
	t4	10.3±1.9	39.95±1.5	31.6±1.9	14.15±7
	t8	7.88±1.7	34.75±3.9	37.05±1.2	16.25±6.8

Reproduced from Mónaco, 1996, PhD thesis, Universidad Nacional de Tucuman, Argentina.

Carbon catabolite repression regulates several genes that code for enzymes of the Embden-Meyerhoff, TCA cycle, gluconeogenesis, and oxidative phosphorylation pathways, the expression of which is controlled by several *cis*- or *trans*-acting regulatory genes (Fig. 4.7) (Entian and Zimmermann, 1982; Gancedo, 1992; Aon and Cortassa, 1995, 1997). Some experimental evidence showed that invertase, respiratory enzymes, and FBPase, are catabolite repressed in cell division cycle (*cdc*) mutants of *S. cerevisiae* (Mónaco *et al.*, 1995; Mónaco, 1996). Furthermore, glucose repression appears to be related with the carbon and energetic uncoupling exhibited by *cdc* mutants at the restrictive temperature, concomitantly with the arrest of cell proliferation, increased ethanolic fermentation, and decreased mitochondrial biogenesis (Mónaco *et al.*, 1995; Aon *et al.*, 1995; Genta *et al.*, 1995; Aon and Cortassa, 1995). Thus, cell cycle-related genes seem to be affected by catabolite repression.

When the same problem was approached from catabolite repression-related genes, similar results were observed. We focused our studies on the effect of catabolite repression-related genes on yeast cell cycle, through disruption of *SNF*1, *SNF*4, and *MIG*1. The two former genes code for proteins involved in catabolite derepression, and the product of the latter has been shown to be involved in glucose repression (Fig. 4.8) (Zimmermann *et al.*, 1977; Celenza and Carlson, 1986; Entian and Zimmermann, 1982; Schuller and Entian, 1987; Gancedo, 1992). It was shown that the onset of fermentative metabolism as well as cell cycle behavior either at cell population or molecular levels in aerobic glucose-limited chemostat cultures, are dependent upon properties linked to the genes studied, known to be involved in the regulation of catabolite (de)repression (Fig. 7.3) (Aon and Cortassa, 1998).

The *snf*4, and to a lesser extent, the *snf*1 mutants exhibited not only effects at the level of the G1 length but also at the differential cell cycle length exhibited by parent and daughter cells in chemostat cultures (Aon and Cortassa, 1998, 1999). In batch cultures, *snf*1 and *snf*4 disruptants accumulated to an extent of 70% or 80% on ethanol, whereas cells arrested randomly when transferred to glycerol, i.e. the percentage of cells in G1 did not evolve after 24 h of exposure to glycerol (Aon and Cortassa, 1998). The level of expression of a *CDC*28-*lacZ* fusion gene in *snf*1 and *snf*4 mutants was two- to three-fold lower than the wild type, and consistently lower on ethanol or glycerol. These mutants are unable to grow on glycerol or ethanol because they do not express gluconeogenic enzymes (Entian and Zimmermann, 1982; Cortassa and Aon, 1998).

Protein Production as a Function of Growth Rate

Growth rate and length of cell cycle phases are strongly related (Fig. 3.6). Increasing evidence suggests that cell-cycle-regulated gene expression plays a crucial role in cell cycle control (Johnston, 1992). Apparently, control of the major cell cycle transitions at Start and at G2-M is exerted principally by post-translational processes, i.e. protein phosphorylation, "checkpoints" that ensure the successful completion of sequential steps in the cell cycle (Hartwell and Weinert, 1989), and cell-cycle-regulated genes. Thus, it is not surprising that production rates of many homologous proteins change as a function of growth rate, μ. Moreover, transcription of malate dehydrogenase and citrate synthase genes of *E. coli*, are inversely proportional to the cell growth rate (Park *et al.*, 1994, 1995) which was varied through different carbon sources or culture conditions such as aerobiosis and anaerobiosis. A temperature-sensitive CHO cell line showed three to four times higher production of a tissue inhibitor of metalloproteinases when growth arrest was induced by a temperature shift to 39°C (Fusseneger *et al.*, 1999).

The implications of the relationship between growth rate, cell cycle regulation, and gene expression, for the fundamental understanding of global regulatory properties of cell function have already been addressed in the previous section. In practical terms, since a continuously decreasing growth rate is a common characteristic of large-scale fed-batch processes, it is relevant to know the relationship between μ and the specific rate of product formation, q_p (Hensing, 1995; see also Chapter 3) (Fig. 7.4).

Inulin is a polyfructan consisting of a linear β-1,2- linked polyfructose chain with a terminal glucose unit at the reducing end. It is found as a reserve carbohydrate in various plants of the Compositae and Graminieae family (Hensing, 1995). The inulinase synthesis as a heterologous protein has been studied in *Kluyveromyces marxianus* under the control of the inulinase promoter (Rouwenhorst *et al.*, 1988). It has been shown that the total amount of inulinase in chemostat cultures of *K. marxianus* is dependent on the dilution rate and the carbon source on which the yeast is grown. Particularly important is the fact that the major amounts of activity are measured at low growth rates under chemostat cultivation on mineral medium (Rouwenhorst *et al.*, 1988).

The situation described in the precedent paragraph seems also to apply to the heterologous expression of an α-galactosidase integrated to the chromosome of *S. cerevisiae* under the control of the *GAL7* promoter (Giuseppin *et al.*, 1993). Production by *S. cerevisiae* of heterologous α-galactosidase was studied quantitatively in chemostat cultures grown with mixtures of glucose and galactose (Giuseppin *et al.*, 1993). The specific productivity, q_p, of the protein increased up to a dilution rate, $D=0.2$ h^{-1}. A maximal q_p of approximately 4.5 g kg^{-1} h^{-1} was attained. At lower growth rates ($D=0.05$ h^{-1}), q_p was still high 3 g kg^{-1} h^{-1} (Giuseppin *et al.*, 1993; see also Hensing *et al.*, 1995). The ratio between intra- and extracellular α-galactosidase was strongly dependent on the growth rate. At low growth rate ($D=0.05$ h^{-1}) about 50% of the α-galactosidase was secreted, increasing still to > 75% at a $D=0.1$ h^{-1} (Giuseppin *et al.*, 1993).

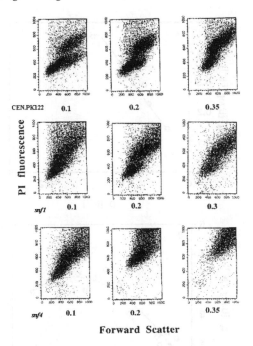

Forward Scatter

Figure 7.3. DNA content as a function of the forward scatter of *S. cerevisiae* wild type CEN.PK 122 (top), and *snf*1 (middle) and *snf*4 (bottom) mutants, at growth rates corresponding to respiratory (D=0.1 h^{-1}), around the onset of ethanolic fermentation (D=0.2 h^{-1}), and the respiro-fermentative (D=0.3-0.35 h^{-1}) regimes of glucose breakdown, in aerobic glucose-limited continuous cultures. At each steady state, cells were analyzed for DNA content by flow cytometry following staining with propidium iodide (PI) using a Becton Dickinson FACSCalibur, as described in Aon and Cortassa (1998). Plotted are 20,000 events (each one of the points represented) whose statistical analysis was performed with the software Cell Quest in a Macintosh Quadra 650 coupled to the flow cytometer (Aon and Cortassa, 1998). The forward light scattering in flow cytometry analysis is proportional to the cross-sectional area of the cells reflecting their size (Lew *et al.*, 1992) and is directly related to the cell volume (Münch *et al.*, 1992). The cloud of points at the bottom of the plot of the wild type, corresponds to cells in G1 whereas the cloud on top coincides with S and G2 cells. As can be noticed in the plots, the DNA content distribution displaces toward larger cells concomitant to the increase in growth rates; the skewness being remarkable for the *snf*4 mutant (bottom panels). It is known that enlarged cells show higher contributions of autofluorescence and that mitochondrial DNA tends to skew populations to higher fluorescence (Richardson *et al.*, 1992). Additional experimental evidence, obtained from plots of cell number as a function of PI fluorescence and fluorescence microscopy, showed displacement of the G2/M peak and enrichment of *snf*4 cells first in S and then in G2/M phases suggesting lengthened S and then G2 phases (Aon and Cortassa, 1998). The distribution of the *snf*1 mutant (middle panels) was similar to the wild type although the separation between cells containing either two-fold (2C) or four-fold (4C) the DNA genomic content was less sharp as in the wild type (top panels).

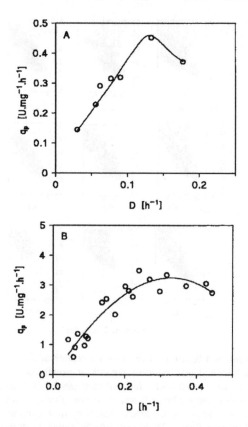

Figure 7.4. Relationship between growth rate and specific product formation rate of two homologous proteins produced by yeast. (A) Production of intracellular alcohol oxidase by *Hansenula polymorpha*, grown in methanol-limited chemostat cultures. (B) Production of extracellular inulinase by *Kluyveromyces marxianus* in sucrose-limited chemostat cultures (Reproduced from Hensing, M.C.M., 1995, PhD thesis Technical University of Delft, The Netherlands).

The Selective Functioning of Whole Metabolic Pathways Is Permissive for Differentiation

Gluconeogenesis is permissive for sporulation in S. cerevisiae. As two faces of the same coin, the metabolic status of yeast cells allow or preclude sporulation. On the one hand, we see the ability of cells to derepress gluconeogenic enzymes, related to the permissive role of gluconeogenesis; and on the other hand, catabolite repression that prevents the differentiation process (reviewed in Cortassa *et al.*, 2000).

The inhibition of sporulation by fermentable carbon sources such as glucose or fructose, has been attributed to the repression of respiration induced by catabolite (Miller, 1989). Existing experimental evidence suggests that the expression of *IME*1 is sensitive to glucose-induced repression of respiration (Treinin and Simchem, 1993). That the product of *IME1* was absent in the presence of glucose, but present in cells growing on acetate, also suggests that this gene is subjected to glucose repression (Kassir *et al.*, 1988).

The *SNF*1 gene codes for a kinase which is involved in the derepression mechanism of gluconeogenic and glyoxylate cycle enzymes (Fig. 4.8); a functional *SNF*1 gene is essential for derepression of glucose-repressed genes (Celenza and Carlson, 1986), and has been implied in regulation of *de novo* fatty acid synthesis through phosphorylation and inactivation of yeast acetyl-CoA carboxylase *in vitro* and *in vivo* (Woods *et al.*, 1994) and glycogen accumulation under nutritional stress (Hardy *et al.*, 1994). Compelling evidence supports the fact that Snf1 is involved in *IME*1 and *IME*2 induction during sporulation. The effect of deleting *SNF*1 is similar to that of adding glucose to a sporulating culture (Honigberg and Lee, 1998), and triggers pleiotropic changes of yeast metabolic status in continuous cultures (Cortassa and Aon, 1998; Aon and Cortassa, 1998). Particularly important was the accumulation of intracellular G6P and ATP depletion shown by the *snf*1 mutant by comparison with the wild type (Cortassa and Aon, 1997). The altered energetic status shown by *snf*1 mutants was observed despite the regime of glucose breakdown (i.e. respiratory or respiro-fermentative) exhibited by the cells in carbon-limited chemostat cultures. At this stage we should also stress that *snf*1 mutants impaired in the derepression of gluconeogenic or glyoxylate cycle enzymes do not sporulate. The latter effects were similar to those exhibited by yeast cells after treatment with the glucose analog, 2-deoxyglucose (2DG). Increasing concentrations of 2DG progressively inhibited sporulation (Aon, J.C. *et al.*, 1997; Cortassa *et al.*, 2000). This catabolite repressor elicited a dramatic inhibition of sporulation efficiency along with a decrease in the rate of appearance of asci (Aon, J.C. *et al.*, 1997). Global anabolic and catabolic fluxes were measured, and fluxes through glyoxylate and gluconeogenic pathways were found to change proportionally to the success of sporulation. Sporulation was inhibited probably due to repression of anabolic pathways and consequently growth restriction.

Particularly, an accumulation of intracellular G6P and ATP depletion with respect to the wild type, were shown to exist in the *snf*1 mutant (Cortassa and Aon, 1997). Indeed, the latter result mimics the effect of 2DG referred above. Interestingly, it has been suggested that the depletion of cellular ATP by environmental stress (or glucose starvation in yeast) and the increase of the AMP:ATP-- ratio could be involved in the activation of the *snf*1 complex which in turn derepress glucose-repressed genes (Hardie *et al.*, 1998; Halford *et al.*, 1999).

Bibliography

Abe, K. and Higuchi, T. (1998) Selective fermentation of xylose by a mutant of *Tetragenococcus halophila* defective in phosphoenolpyruvate: mannose phosphotransferase, phosphofructokinase, and glucokinase. *Biosci. Biotechnol. Biochem.* 62, 2062-2064.

Abraham, R.H. and Shaw, Ch.D. (1987) Dynamics: A visual introduction. In *Self-Organizing Systems. The Emergence of Order*, (Yates, F.E., Garfinkel, A., Walter, D. O., and Yates, G. B., eds.), pp. 543-597. Plenum Press, New York.

Abraham, R.H. (1987) Dynamics and self-organization. In *Self-Organizing Systems. The Emergence of Order* (Yates, F.E., Garfinkel, A., Walter, D.O., and Yates, G.B., eds.), pp. 599-613. Plenum Press, New York.

Aiba, S., Humphrey, A.E. and Millis, N.F. (1973) *Biochemical Engineering*, 2nd edn. Academic Press.

Alberts, B., Bray, D., Lewis, J., Raff, M., Roberts, K. and Watson, J.D. (1989) *Molecular Biology of the Cell*, Garland Publishing, New York.

Altaras, N.E. and Cameron, D.C. (1999) Metabolic engineering of a 1,2-propanediol pathway in *Escherichia coli*. *Appl. Environm. Microbiol.* 65, 1180-1185.

Anderson, A.J. and Dawes, E.A. (1990) Occurrence, metabolism, metabolic role, and industrial uses of bacterial polyhydroxyalkanoates. *Microbiol. Rev.*, 54, 450-472.

Andrews, J.F. (1968) A mathematical model for the continuous culture of microorganisms utilizing inhibitory substances. *Biotech. Bioengin.* 10, 707-723.

Aon, M.A., Cortassa, S., Westerhoff, H.V., Berden, J. A., van Spronsen, E. and van Dam, K. (1991) Dynamic regulation of yeast glycolytic oscillations by mitochondrial functions. *J. Cell Sci.*, 99, 325-334.

Aon, M.A. and Cortassa, S. (1991) Thermodynamic evaluation of energy metabolism in mixed substrate catabolism. Modeling studies of stationary and oscillatory states. *Biotech. Bioengin.* 37, 197-204.

Aon, M.A. and Cortassa, S. (1993) An allometric interpretation of the spatio-temporal organisation of molecular and cellular processes. *Mol. Cell. Biochem.* 120, 1-14.

Aon, M.A., Mónaco, M.E. and Cortassa, S. (1995) Carbon and energetic uncoupling are associated with block of division at different stages of the cell cycle in several cdc mutants of *Saccharomyces cerevisiae*. *Exp. Cell Res.* 217, 42-51.

Aon, M.A. and S. Cortassa (1995) Cell growth and differentiation from the perspective of dynamical organization of cellular and subcellular processes. *Prog. Biophys. Mol. Biol.* 64, 55-79.

Aon, J.C. and Cortassa, S. (1996) Metabolic rates during sporulation of *Saccharomyces cerevisiae* on acetate. *Antonie van Leeuwenhoek* 69, 257-265.

Aon, J.C., Rapisarda, V.A. and Cortassa, S. (1996) Metabolic rates regulate the success of sporulation in *Saccharomyces cerevisiae*. *Exp. Cell Res.* 222, 157-162.

Aon, J.C. (1996) Bioenergética y regulación metabólica durante la esporulación en *Saccharomyces cerevisiae*. PhD Thesis, Universidad Nacional de Tucumán, Argentina.

Aon, M.A. and Mónaco, M.E. (1996) Catabolism regulates carbon and energetic imbalance during arrest of cell proliferation in *Saccharomyces cerevisiae. Biochim. Biophys. Acta* 9, 237 (Abstr.).

Aon M.A., Cortassa S and Cáceres A. (1996) Models of cytoplasmic structure and function. In *Computation in Cellular and Molecular Biological Systems* (Cuthbertson, R., Holcombe, M. and Paton, R., eds.), pp.195-207. World Scientific, London.

Aon, J.C., Aon, M.A., Spencer, J.F.T. and Cortassa, S. (1997) Modulation of sporulation and metabolic fluxes in *Saccharomyces cerevisiae* by 2 Deoxy glucose. *Antonie van Leeuwenhoek* 72, 283-290.

Aon, M.A. and Cortassa, S. (1997) *Dynamic Biological Organization*. Fundamentals as applied to Cellular Systems. Chapman & Hall, London.

Aon, M.A. and Cortassa, S. (1998) Catabolite repression mutants of *Saccharomyces cerevisiae* show altered fermentative metabolism as well as cell cycle behavior in glucose-limited chemostat cultures. *Biotechnol. Bioengin.* 59, 203-213.

Aon, M.A. and Cortassa, S. (1999) Quantitation of the effects of disruption of catabolite (de)repression genes on the cell cycle behavior of *Saccharomyces cerevisiae. Curr. Microbiol.* 38, 57-60.

Aon, M.A., Cortassa, S., Gomez Casati, D.F. and Iglesias, A.A. (2000a) Effects of stress on cellular infrastructure and metabolic organization in plant cells. *Int. Rev. Cytol.* 194, 239-273.

Aon, M.A., Cortassa, S. and Lloyd, D. (2000b) Fractal space and chaotic dynamics in biochemistry: Simplicity underlies complexity. *Cell Biol. Int.* 24, 581-587.

Aon, M.A., Gomez-Casati, D.F., Iglesias, A.A. and Cortassa, S. (2001) Ultrasensitivity in (supra)molecularly organized and crowded environments. *Cell Biol. Int.* 25, 1091-1099.

Aon, J.C. and Cortassa, S. (2001) Involvement of central carbon and nitrogen metabolic pathways in the triggering of ethanolic fermentation in aerobic chemostat cultures of *Saccharomyces cerevisiae. Metab. Engin.* 3, 250-264.

apRees, T. and Hill, S.A. (1994) Metabolic control analysis of plant metabolism. *Plant Cell Environm.* 17, 587-599.

Aris, P. (1969) *Elementary Chemical Reactor Analysis*. Prentice Hall Inc. Englewood Cliffs, New Jersey.

Aristidou, A.A., San, K.Y. and Bennett, G.N. (1995) Metabolic engineering of *Escherichia coli* to enhance recombinant protein production through acetate reduction. *Biotechnol. Prog.* 11, 475-478.

Auberson, L.C.M., Kanbier, T. and von Stockar, U. (1993) Monitoring synchronized yeast cultures by calorimetry. *J. Biotech.* 29, 205-215.

Bailey, J.F. and D.F. Ollis (1977) *Biochemical Engineering Fundamentals*. McGraw-Hill, New York.

Bailey, J.E., D.D. Axe, Doran, P.M., Galazzo, J.L., Reardon, K.F., Seressiotis, A. and Shanks, J.V. (1987) Redirection of cellular metabolism: Analysis and synthesis. *Ann. New York Acad. Sci.*, 506, 1-23.

Bailey, J.E. (1991) Toward a science of metabolic engineering. *Science* 252, 1668-1675.

Bailey, J.E. (1998) Mathematical modeling and analysis in biochemical engineering: Past accomplishments and future opportunities. *Biotechnol. Prog.* 14, 8-20.

Ball, S.G. (1995) Recent views on the biosynthesis of the starch granule. *Trends Glycosci. Glycotechnol.* 7, 405-415.

Barford, J.P., Pamment, N.B. and Hall, R.J. (1982) Lag phases and transients pp. 56-89. In Microbial Population Dynamics, (Bazin, M.J., ed.). CRC Press, Boca Raton, Florida, pp. 56-89.

Bartels, D. (2001) Targeting detoxification pathways: An efficient approach to obtain plants with multiple stress tolerance. *Trends Plant Sci.* 6, 284-286.

Bauchop, T. and Elsden, S.R. (1960) The growth of microorganisms in relation to their energy supply. *J. Gen. Microbiol.* 23, 457-469.

Beachy, R.N. (1997) Plant Biotechnology: The now and then of plant biotechnology. *Curr. Opin. Biotechnol.* 8, 187-188.

Bell, S.L., Bebbington, C., Scott, M.F., Wardell, J.N., Spier, R.E., Bushell, M.E. and Sanders, P.G. (1995) Genetic engineering of hybridoma glutamine metabolism. *Enz. Microb. Technol.* 17, 98-106.

Bertalanffy L. Von (1950) The theory of open systems in physics and biology. *Science* 111, 23-29.

Bhalla, U.S. and Iyengar, R. (1999) Emergent properties of networks of biological signalling pathways. *Science* 283, 381-387.

Birch, R.G. (1997) Plant Transformation: Problems and strategies for practical application. *Annu. Rev. Plant Physiol. Plant Mol. Biol.* 48, 297-326.

Bonarius, H.P.J., Hatzimanikatis, V., Meesters, K.P.H., de Gooijer, C.D., Schmid, G. and Tramper, J. (1996) Metabolic flux analysis of hybridoma cells in different culture media using mass balances. *Biotechnol. Bioengin.* 50, 299-318.

Bray, D. (1995) Protein molecules as computational elements in living cells. Nature 376, 307-312.

Brazil, G.M., Kenefick, L., Callanan, M., Haro, A., de Lorenzo, V., Dowling, D.N. and O'Gara, F. (1995) Construction of a rhizosphere pseudomonad with potential to degrade polychlorinated biphenyls and detection of bph gene expression in the rhizosphere. *Appl. Environm. Microbiol.* 61, 1946-1952.

Brim, H., McFarlan S.C., Fredrickson, J.K., Minton, K.W., Zhai, M. Wackett, L.P. and Daly, M.J. (2000) Engineering *Deinococcus radiodurans* for metal remediation in radioactive mixed waste environments. *Nature Biotechnol.* 18, 85-90

Brower, J.R. and Leegood, R.C. (1997) Photosynthesis, in: *Plant Biochemistry* (Dey, P.M., and Harborne, J.B., eds.). Chapter 2, pp. 49-110. Academic Press, San Diego.

Brown, G.C. (1992) Control of respiration and ATP synthesis in mammalian mitochondria and cells. *Biochem. J.* 284, 1-13.

Bruinenberg, P.M., van Dijken, J.P. and Scheffers, W.A. (1983) A theoretical analysis of NADPH production and consumption in yeast. *J. Gen. Microbiol.* 129, 953-964.

Burns, N., Grimwade, B., Ross-Macdonald, P.B., Choi, E.-Y., Finberg, K., Roeder, G.S. and Snyder, M. (1994) Large-scale analysis of gene expression, protein localization, and gene disruption in *Saccharomyces cerevisiae. Genes & Developm.* 8, 1087-1105.

Caddick, M.X., Greenland, A.J., Jepson, I., Krause, K.-P., Qu, N., Riddell, K.V., Salter, M.G., Schuch, W., Sonnewald, U. and Tomsett, A.B. (1998) An ethanol inducible gene switch for plants used to manipulate carbon metabolism. *Nature Biotechnol.* 16, 177-180.

Cameron, D.C. and Tong, I.T. (1993) Cellular and metabolic engineering. *Appl. Biochem. Biotechnol.* 38, 105-140.

Cameron, D.C. and Chaplen, F.W.R. (1997) Developments in metabolic engineering. *Curr. Opin. Biotechnol.* 8, 175-180.

Cameron, D.C., Altaras, N.E., Hoffman, M.L. and Shaw, A.J. (1998) Metabolic engineering of propanediol pathways. *Biotechnol. Prog.* 14, 116-125.

Celenza, J.L. and Carlson, M. (1986) A yeast gene that is essential for release from glucose repression encodes a protein kinase. *Science* 233, 1175-1180.

Celenza, J.L. and Carlson, M. (1989) Mutational analysis of *Saccharomyces cerevisiae* *SNF*1 protein kinase and evidence for functional interaction with the *SNF*4 protein. *Mol. Cell. Biol.* 5034-5044.

Chance, B., Pye, K. and Higgins, J. (1967) Waveform generation by enzymatic oscillators. *IEEE Spectrum* 4, 79-86.

Christou, P. (1992) Genetic transformation of crop plants using microprojectile bombardment. *Plant* J. 2, 275-281.

Clegg, J.S. (1984) Properties and metabolism of the aqueous cytoplasm and its boundaries. *Am. J. Physiol.* 246, R133-R151.

Clegg, J.S. (1991) Metabolic organization and the ultrastructure of animal cells. *Biochem. Soc. Trans.* 19, 985-991.

Cleveland, D.W. (1988) Autoregulated instability of tubulin mRNAs: a novel eukaryotic regulatory mechanism. *TIBS* 13, 339-343.

Collins, G.B. and Shepherd, R.J., Eds. (1996) *Engineering Plants for Commercial Products and Applications*, 183 pp., Annals of the New York Academy of Sciences, Vol. 792.

Cooney, C.L., Wang, H.Y. and Wang, D.I.C. (1977) Computer-aided material balancing for the prediction of fermentation parameters. *Biotech. Bioengin.* 19, 55-67.

Cooper, S. (1991) *Bacterial Growth and Division*. Academic Press, San Diego.

Cornish-Bowden, A. (1989) Metabolic Control Theory and Biochemical Systems Theory: different objectives, different assumptions, different results. *J. Theor. Biol.* 136, 365-377.

Cornish-Bowden, A. (1995) Kinetics of multi-enzyme systems. In *Biotechnology* (Rehm, H.J., Reed, G., Puhler, A. and Stadler, P., eds.). *Enzymes, Biomass, Food and Feed* (Reed, G. and Nagodawithana, T.W., eds.) vol. 9, pp. 122-136. VCH, Weinheim.

Cortassa, S., Aon, M.A. and Thomas, D. (1990) Thermodynamic and kinetic studies in a stoichiometric model of energetic metabolism under starvation conditions. *FEMS Microbiol. Lett.* 66, 249-256.

Cortassa, S., Aon, M.A. and Westerhoff, H.V. (1991) Linear non equilibrium thermodynamics describes the dynamics of an autocatalytic system. *Biophys. J.* 60, 794-803.

Cortassa, S. and Aon, M.A. (1994a) Metabolic control analysis of glycolysis and branching to ethanol production in chemostat cultures of *Saccharomyces cerevisiae* under carbon, nitrogen, or phosphate limitations. *Enz. Microb. Technol.* 16, 761-770.

Cortassa, S. and Aon, M.A. (1994b) Spatio-temporal regulation of glycolysis and oxidative phosphorylation *in vivo* in tumour and yeast cells. *Cell Biol. Int.* 89, 687-714.

Cortassa, S., Aon, J.C. and Aon, M.A. (1995) Fluxes of carbon, phosphorylation and redox intermediates during growth of *Saccharomyces cerevisiae* on different carbon sources. *Biotech. Bioengin.* 47, 193-208.

Cortassa, S. and Aon, M.A. (1996) Entrainment of enzymatic activity by the dynamics of cytoskeleton. In *Biothermokinetics* (Westerhoff, H.V. and Snoep, J. eds). Amsterdam.

Cortassa, S. and Aon, M.A. (1997) Distributed control of the glycolytic flux in wild-type cells and catabolite repression mutants of *Saccharomyces cerevisiae* growing in carbon-limited chemostat cultures. *Enz. Microb. Technol.* 21, 596-602.

Cortassa, S. and Aon, M.A. (1998) The onset of fermentative metabolism in continuous cultures depends on the catabolite repression properties of *Saccharomyces cerevisiae*. *Enz. Microb. Technol.* 22, 705-712.

Cortassa, S., Aon, J.C., Aon, M.A. and Spencer, J.F.T. (2000) Dynamics of metabolism and its interactions with gene expression during sporulation in *Saccharomyces cerevisiae*. *Adv. Microb. Physiol.* 43, 75-115.

Cortassa, S., Gomez-Casati, D.F., Iglesias, A.A. and Aon, M.A. (2001) Mathematical modeling of the ultrasensitive behavior of the glycogen synthetic pathway in cyanobacteria (submitted).

Coshigano, P.W., Miller, S.M. and Magasanik, B. (1991) Physiological and genetic analysis of the carbon regulation of the NAD-dependent glutamate dehydrogenase of *Saccharomyces cerevisiae. Mol. Cell. Biol.* 11, 4455-4465.

Covert, M.W., Schilling, Ch.H., Famili, I., Edwards, J.S., Goryanin, I.I., Selkov, E. and Palsson, B.O. (2001) Metabolic modeling of microbial strains *in silico. TIBS* 26, 179-186.

Cremer, J., Eggeling, L. and Sahm, H. (1991) Control of the lysine biosynthesis sequence in *Corynebacterium glutamicum* as analyzed by overexpression of the individual corresponding genes. *Appl. Environm. Microbiol.* 57, 1746-1752.

Curds, C.R. and Cockburn, A. (1971) Continuous monoxenic culture of *Tetrahymena pyriformis. J. Gen. Microbiol.* 66, 95-108.

Dang, V.D., Bohn, C., Bolotin-Fukuhara, M. and Daignan-Fornier, B. (1996) The CCAAT box-binding stimulates ammonium assimilation in *Saccharomyces cerevisiae*, defining a new cross-pathway regulation between nitrogen and carbon metabolisms. *J. Bacteriol.* 178, 1842-1849.

Daniell, H., Datta, R., Varma, S. Gray, S. and Lee, S.-B. (1998) Containment of herbicide resistance through genetic engineering of the chloroplast genome. *Nature Biotechnol.* 16, 345-348.

Dave, E., Guest, J.R. and Attwood, M.M. (1995) Metabolic engineering in *Escherichia coli*: lowering the lipoyl domain content of the pyruvate dehydrogenase complex adversely affects the growth rate and yield. *Microbiology* 141, 1839-1849.

Davey, H.M., Davey, C.L., Woodward, A.M., Edmonds, A.N., Lee, A.W. and Kell, D.B. (1996) Oscillatory stochastic and chaotic growth rate fluctuations in permittistatically controlled yeast cultures. *BioSystems* 39, 43-61.

Dawson, P.S.S. (1985) Continuous cultivation of microorganisms. *CRC Crit. Rev. Biotechnol.* 2, 315-374.

Dean, A.C. and Moss, D.A. (1970) Interaction of nalidixic acid with *Klebsiella (Aerobacter) aerogenes* growing in continuous culture. *Chem. Biol. Interact.* 2, 281-296.

Deanda, K., Zhang, M., Eddy, C. and Picataggio, S. (1996) Development of an arabinose-fermenting *Zymomonas mobilis* strain by metabolic pathway engineering. *Appl. Environm. Microbiol.* 62, 4465-4470.

de Groot, M.J.A., Bundock, P., Hooykaas, P.J.J. and Beijersbergen, A.G.M. (1998) *Agrobacterium tumefaciens*-mediated transformation of filamentous fungi. *Nature Biotechnol.* 16, 839-842.

Degn, H. and Harrison, D.E.F. (1969) Theory of oscillations of respiration rate in continuous culture of *Klebsiella aerogenes. J. Theor. Biol.* 22, 238-248.

De Jong-Gubbels, P., Vanrolleghem, P., Heijnen, S., Van Dijken, J.P. and Pronk, J.T. (1995) Regulation of carbon metabolism in chemostat cultures of *Saccharomyces cerevisiae* grown on mixtures of glucose and ethanol. *Yeast* 11, 407-418.

Della Penna, D. (2001) Plant metabolic engineering. *Plant Physiol.* 125, 160-163.

Demple, B. (1998) A bridge to control. *Science* 279, 1655-1656.

Dixon, R.A., Ed. (1985) *Plant Cell Culture: A Practical Approach*, 232 pp., IRL Press, Oxford.

Dixon, R.A., Lamb, C.J.; Masoud, S.; Sewalt, V.J.H. and Paiva, N.L. (1996) Metabolic Engineering: prospects for crop improvement through the genetic manipulation of phelylpropanoid biosynthesis and defence response—a review. *Gene* 179, 61-71.

Doedle, E. (1986) *AUTO Manual*. California Institute of Technology, Pasadena.

Domach, M.M. and Shuler, M.L. (1984) A finite representation model for an asynchronous culture of *E. coli. Biotech. Bioengin.* 26, 877-884.

Domach, M.M., Leung, S.K., Cahn, R.E., Cocks, G.G. and Shuler, M.L. (1984) Computer model for glucose limited growth of a single copy of *E. coli* B/r-A *Biotech. Bioengin.* 26, 203-216.

Doran, P.M. (1995) *Bioprocess Engineering Principles*. Academic Press, London.

Duboc, P., Marison, I. and von Stockar, U. (1996) Physiology of *Saccharomyces cerevisiae* during cell cycle oscillations. *J. Biotechnol.* 51, 57-72.

Dunbar, J., Campbell, S.L., Banks, D.J. and Warren, D.R. (1998) Metabolic aspects of a commercial continuous fermentation system. *Aust. N. Z. Sect. Proc. 20th Conv. Brisbane*, 151-158.

Duport, C., Spagnoli, R., Degryse, E. and Pompon, D. (1998) Self-sufficient biosynthesis of pregnenolone and progesterone in engineered yeast. *Nature Biotechnol.* 16, 186-189.

Dworkin, M.B. and Dworkin-Rastl, E. (1989a) Metabolic regulation during early frog development: Glycogenic flux in *Xenopus* oocytes, eggs, and embryos. *Dev. Biol,* 132, 512-523.

Dworkin, M.B. and Dworkin-Rastl, E. (1989b) Metabolic regulation during early frog development: Flow of glycolytic carbon into phospholipids in *Xenopus* oocytes and fertilized eggs. *Dev. Biol.* 132, 524-528.

Dworkin, M.B. and Dworkin-Rastl, E. (1991) Carbon metabolism in early amphibian embryos. *TIBS* 16, 229-234.

Dykhuizen, D.E., Dean, A.M. and Hartl, D.L. (1987) Metabolic flux and fitness. *Genetics* 115, 25-31.

Edmunds, L.N. Jr. (1988) *Cellular and Molecular Bases of Biological Clocks*. Models and mechanisms for circadian timekeeping. Springer, New York.

Edwards, G.E. (1999) Tuning up crop photosynthesis. *Nature Biotechnol.* 17, 22-23.

Edwards, J.S., Ibarra, R.U. and Palsson, B.O. (2001) *In silico* predictions of *Escherichia coli* metabolic capabilities are consistent with experimental data. *Nature Biotechnol.* 19, 125-130.

Eggeling, L. and Sahm, H. (1999) Amino acid production: Principles of metabolic engineering. In *Metabolic Engineering* (Lee, S.Y and Papoutsakis, E.T., eds.), pp. 153-176. Marcel Dekker, New York.

Emes, M.J., Bowcher, C.G., Debnam, P.M., Dennis, D.T., Hanke, G., Rawsthorne, S. and Tetlow, I.J. (1999) Implications of inter- and intracellular compartmentation for the movement of metabolites in plant cells. In *Plant Carbohydrate Biochemistry* (Bryant, J.A., Burnell, M.M., and Kruger, N.J., eds.). Chap. 16, pp. 231-244. BIOS Scientific Publishers Ltd., Oxford.

Emmerling, M., Bailey, J.E. and Sauer, U. (1999) Glucose catabolism of *Escherichia coli* strains with increased activity and altered regulation of key glycolytic enzymes. *Metab. Engin.* 1, 117-127.

Entian, K.D. and Zimmermann, F.K. (1982) New genes involved in carbon catabolite repression and derepression in the yeast *Saccharomyces cerevisiae. J. Bacteriol.* 151, 1123-1128.

Entian, K.D. and Barnett, J.A. (1992) Regulation of sugar utilization *by Saccharomyces cerevisiae. Trends Biochem. Sci.* 17, 506-510.

Erickson, L.E., Minkevich, I.G. and Eroshin, V.K. (1978) Application of mass and energy balance regularities in fermentation. *Biotech. Bioengin.* 20, 1595-1621.

Esener, A.A., Roels, J.A. and Kossen, N.W.F. (1983) Theory and applications of unstructured growth models: Kinetic and energetic aspects. *Biotech. Bioeng.* 25, 2803-2841.

Farmer, W.R. and Liao, J.C. (2001) Precursor balancing for metabolic engineering of lycopene production in *Escherichia coli. Biotechnol. Prog.* 17, 57-61

Farmer, W.R. and Liao, J.C. (1996) Progress in metabolic engineering. *Curr. Op. Biotechnol.* 7, 198-204.

Fell, D.A. (1992) Metabolic Control Analysis: A survey of its theoretical and experimental development. *Biochem. J.* 286, 313-330.

Fell, D.A. (1998) Increasing the flux in metabolic pathways: A metabolic control analysis perspective. *Biotech. Bioengin.* 58, 121-124.

Ferrell, Jr., J.E. (1998) How regulated protein translocation can produce switch-like responses. *Trends Biochem. Sci.* 23, 461-465.

Fiehn, O., Kopka, J., Dörmann, P., Altmann, T., Trethewey, R.N. and Willmitzer, L. (2000) Metabolite profiling for plant functional genomics. *Nature Biotechnol.* 18, 1157-1161.

Fischer, R., Drossard, J., Commandeur, U., Schillberg, S. and Emans, N. (1999a) Towards molecular farming in the future: moving form diagnostic and antibody production in microbes to plants. *Biotechnol. Appl. Biochem.* 30, 101-108.

Fischer, R., Emans, N., Schuster, F., Hellwig, S. and Drossard, J. (1999b) Towards Molecular Farming in the future: using plant-cell-suspension cultures as bioreactors. *Biotechnol. Appl. Biochem.* 30, 109-112.

Fischer, R., Vaquero-Martin, C., Sack, M., Drossard, J., Emans, N. and Commandeur, U. (1999c) Towards molecular farming in the future: Transgenic protein expression in plants. *Biotechnol. Appl. Biochem.* 30, 113-116.

Flores, N., Xiao, J., Berry, A., Bolivar, F. and Valle, F. (1996) Pathway engineering for the production of aromatic compounds in *Escherichia coli*. *Nature Biotechnol.* 14, 620-623.

Flores-Samaniego, B., Olivera, H. and Gonzalez, A. (1993) Glutamine synthesis is a regulatory signal controling glucose catabolism in *Saccharomyces cerevisiae*. *J. Bacteriol.* 175, 7705-7706.

Forsburg, S.L. and Nurse, P. (1991) Cell cycle regulation in the yeasts *Saccharomyces cerevisiae* and *Schizosaccharomyces pombe*. *Annu. Rev. Cell Biol.* 7, 227-256.

Fraenkel, D.G. (1992) Genetics and intermediary metabolism. *Annu. Rev. Genet.* 26, 159-177.

French, C.E., Rossar, S.J., Davies, G.J., Nickling, S. and Bruce, N.C. (1999) Biodegradation of explosives by transgenic plants expressing pentaerythritol tetranitrate reductase. *Nature Biotechnol.* 17, 491-494

Fussenegger, M., Schlatter, S., Datwyler, D., Mazur, X. and Bailey, J.E. (1998) Controlled proliferation by multigene metabolic engineering enhances the productivity of Chinese hamster ovary cells. *Nature Biotechnol.* 16, 468-472.

Fussenegger, M., Sburlati, A. and Bailey, J.E. (1999) Metabolic engineering of mammalian cells. In *Metabolic Engineering* (Lee, S.Y. and Papoutsakis, E.T., eds.), pp. 353-389. Marcel Dekker, New York.

Gad, A.E., Rosenberg, N. and Altman, A. (1990) Liposome-mediated gene delivery into plant cells. *Physiol. Plant.* 79, 177-183.

Gancedo, J.M. (1992) Carbon catabolite repression in yeast. *Eur. J. Biochem.* 206, 297-313.

Gauthier, D.A. and Turpin, D.H. (1997) Interactions between inorganic phosphate (Pi) assimilation, photosynthesis and respiration in the Pi-limited green alga *Selenastrum minutum*. *Plant Cell Environ.* 20, 12-24.

Genta, H.D., Mónaco, M.E. and Aon, M.A. (1995) Decreased mitochondrial biogenesis in temperature-sensitive cell division cycle mutants of *Saccharomyces cerevisiae*. *Curr. Microbiol.* 31, 327-331.

Ghosh, A. and Chance, B. (1964) Oscillations of glycolytic intermediate in yeast cells. *Biochem. Biophys. Res. Commun.* 16, 174-181.

Giuseppin, M.L.F., Almkerk, J.W., Heistek, J.C. and Verrips, C.T. (1993) Comparative study on the production of Guar α-galactosidase by *Saccharomyces cerevisiae* SU50B and *Hansenula polymorpha* 8/2 in continuous cultures. *Appl. Environm. Microbiol.* 59, 52-59.

Giuseppin, M.L.F. and van Riel, N.A.W. (2000) Metabolic modeling of *Saccharomyces cerevisiae* using the optimal control of homeostasis: A cybernetic model definition. *Metab. Engin.* 2, 14-33.

Goel, A., Lee, J., Domach, M.M. and Ataai, M.M. (1995) Suppressed acid formation by cofeeding of glucose and citrate in *Bacillus* cultures: emergence of pyruvate kinase as a potential metabolic engineering site. *Biotechnol Prog* 11(4), 380-385.

Goldbeter, A. and Koshland, Jr., D.E. (1981) An amplified sensitivity arising from covalent modification in biological systems. *Proc. Natl. Acad. Sci. USA* 78, 6840-6844.

Goldbeter, A. (1996) *Biochemical Oscillations and Cellular Rhythms.* Cambridge, University Press.

Gomez Casati, D.F., Aon, M.A. and Iglesias, A.A. (1999) Ultrasensitive glycogen synthesis in *Cyanobacteria. FEBS Lett.* 446, 117-121.

Gomez Casati, D.F., Aon, M.A. and Iglesias, A.A. (2000) Kinetic and structural analysis of the ultrasensitive behavior of cyanobacterial ADPGlucose Pyrophosphorylase. *Biochem. J.* 350, 139-147.

Gomez Casati, D.F., Aon, M.A., Cortassa, S. and Iglesias, A.A. (2001) Measurement of the glycogen synthetic pathway in permeabilized cells of cyanobacteria. *FEMS Microbiol. Lett.* 194, 7-11.

Gommers, P.J.F., van Schie, B.J., van Dijken, J.P. and Kuenen, J.G. (1988) Biochemical limits to microbial growth yields: An analysis of mixed substrate utilization. *Biotech. Bioeng.* 32, 86-94.

Goodenough, U.W. (1992) Green Yeast. *Cell* 70, 533-538.

Gosset, G., Yong-Xiao, J. and Berry, A. (1996) A direct comparison of approaches for increasing carbon flow to aromatic biosynthesis in *Escherichia coli. J. Ind. Microbiol.* 17, 47-52.

Gubb, D. (1993) Genes controling cellular polarity in *Drosophila. J. Cell Sci.* Suppl. 269-273.

Gubler, M., Park, S.M., Jetten, M., Stephanopoulos, G. and Sinskey, A.J. (1994) Effects of phosphoenol pyruvate carboxylase deficiency on metabolism and lysine production in *Corynebacterium glutamicum. Appl. Microbiol. Biotechnol.* 40, 857-863.

Guda, C., Lee, S.B. and Daniell, H. (2000) Stable expression of a biodegradable protein-based polymer in tobacco chloroplasts. *Plant Cell Rep.* 19, 257-262.

Gundersen, G.G. and Cook, T.A. (1999) Microtubules and signal transduction. *Curr. Op. Cell Biol.* 11, 81-94.

Haken, H. (1978) *Synergetics.* Springer-Verlag, Berlin, Heidelberg.

Halford, N.G., Purcell, P.C. and Hardie, D.G. (1999). Is hexokinase really a sugar sensor in plants? *Trends Plant Sci.* 4, 117-120

Hardie, D.G., Carling, D. and Carlson, M. (1998). The AMP-activated/SNF1 protein kinase subfamily: Metabolic sensors of the eukaryotic cell? *Annu. Rev. Biochem.* 67, 821-855.

Hardy, T.A., Huang, D. and Roach, P.J. (1994) Interactions between cAMP-dependent and *SNF1* protein kinases in the control of glycogen accumulation in *Saccharomyces cerevisiae. J. Biol. Chem.* 269, 27907-27913.

Harold, F.M. (1990) To shape a cell: an inquiry into the causes of morphogenesis of microorganisms. *Microbiol. Rev.* 54, 381-431.

Harrison, D.E.F. and Pirt, S.J. (1967) The influence of dissolved oxygen concentration on the respiration and glucose metabolism of *Klebsiella aerogenes* during growth. *J. Gen. Microbiol.* 46, 193-211.

Harrison, D.E.F., MacLennan, D.G. and Pirt, S.J. (1969) *Responses of Bacteria to Dissolved Oxygen Tension.* Fermentation advances. Academic Press, New York.

Harrison, D.E.F. (1970) Undamped oscillations of pyridine nucleotide and oxygen tension in chemostat cultures of *Klebsiella aerogenes. J. Cell Biol.* 45, 514-521.

Harrison, D.E.F. and Loveless, J.E. (1971) Transient responses of facultatively anaerobic bacteria growing in chemostat culture to a change from anaerobic to aerobic conditions. *J. Gen. Microbiol.* 68, 45-52

Harrison, D.E.F. and Topiwala, H.H. (1974) Transient and oscillatory states of continuous culture. In *Adv. Biochem. Engin.* Vol. 3 (Ghose, T.K., Fiechter, A. and Blakebrough, N., eds.) Springer Verlag, Berlin.

Hartwell, L.H. and Weinert, T.A. (1989) Checkpoints: Controls that ensure the order of cell cycle events. *Science* 246, 629-634.

Hartwell, L.H. (1974) *Saccharomyces cerevisiae* cell cycle. *Bacteriol. Rev.* 38, 164-198.

Hartwell, L.H. (1991) Twenty-five years of cell cycle genetics. *Genetics* 129, 975-980.

Hatzimanikatis, V., Floudas, C. and Bailey, J.E. (1996) Analysis and design of metabolic reaction networks via mixed-integer linear optimisation. *AIChE J.* 42, 1277-1292.

Hatzimanikatis, V. and Bailey, J.E. (1997) Effects of spatiotemporal variations on metabolic control: Approximate analysis using (log) linear kinetic models. *Biotech. Bioengin.* 54, 91-104.

Heinrich, R. and Rappoport, T. (1974) A linear steady-state treatment of enzymatic chains. General properties, control and effector strength. *Eur. J. Biochem.*, 42, 89-95.

Heinrich, R., Rappoport, S.M. and Rappoport, T.A. (1977) Metabolic regulation and mathematical models. *Prog. Biophys. Mol. Biol.*, 32, 1-82.

Heldt, H.W., Chon, Ch.J., Maronde, D., Herold, A., Stankovic, Z.S., Walker, D.A., Kraminer, A., Kirk, M.R. and Heber, U. (1977) Role of ortophosphate and other factors in the regulation of starch formation in leaves and isolated chloroplasts. *Plant Physiol.* 59, 1146-1155.

Hensing, M. (1995) Production of extracellular proteins by *Kluyveromyces* yeasts. PhD Thesis, Delft University of Technology, The Netherlands.

Hensing, M.C.M., Bangma, K.A., Raamsdonk, L.M., de Hulster, E., van Dijken, J.P. and Pronk, J.T. (1995) Effects of cultivation conditions on the production of heterologous α-galactosidase by *Kluyveromyces lactis. Appl. Microbiol. Biotechnol.* 43, 58-64.

Herbers, K. and Sonnewald, U. (1996) Manipulating metabolic partitioning in transgenic plants. *Trends Biotechnol.* 14, 198-205.

Herrmann, K.M. and Weaver, L.M. (1999) The shikimate pathway. *Annu. Rev. Plant Physiol. Plant Mol. Biol.* 50, 473-503.

Heyer, A.G., Lloyd, J.R. and Kossmann, J. (1999) Production of modified polymeric carbohydrates. *Curr. Opinion Biotechnol.* 10, 169-174.

Higgins, J. (1965). Dynamics and control in cellular reactions. In *Control of Energy Metabolism* (Chance, B., Estabrook, R.W. and Williamson, J.R., eds.), pp. 13-46. Academic Press, New York.

Higgins, J. (1967) The theory of oscillating reactions. *Ind. Engng. Chem.* 59, 19-62.

Himmelblau, D.M. and Bischoff, K.B. (1968) *Process Analysis and Simulations.* John Wiley & Sons Inc. New York.

Hofmeyr, J.H.S. and Cornish-Bowden, A. (1991) Quantitative assessment of regulation in metabolic systems. *Eur. J. Biochem.* 200, 223-236.

Holmberg, N., Lilius, G., Bailey, J.E. and Bülow, L. (1997) Transgenic tobacco expressing Vitreoscilla hemoglobin exhibits enhanced growth and altered metabolite production. *Nature Biotechnol.* 15, 244-247.

Holms, W.H. (1986) The central metabolic pathways of *Escherichia coli*: Relationships between flux and control at a branch point, efficiency of conversion to biomass and excretion of acetate. *Curr. Top. Cell. Regul.*, 28, 69-105.

Hols, P., Kleerebezem, M., Schanck, A.N., Ferain, T., Hugenholtz, J., Delcour, J. and deVos, W.M. (1999) Conversion of *Lactococcus lactis* from homolactic to homoalanine fermentation through metabolic engineering. *Nature Biotechnol.* 17, 588-592.

Honigberg, S.M. and Lee, R.H. (1998) Snf1 kinase connects nutritional pathways controling meiosis in *Saccharomyces cerevisiae. Mol. Cell. Biol.* 18, 4548-4555.

Hooykaas, P.J.J. and Schilperoort, R.A. (1992) *Agrobacterium* and plant genetic engineering. *Plant Mol. Biol.* 19, 15-38.

Hu, W-J, Harding, S.A., Lung, J., Popko, J.L., Ralph, J., Stokke, D.D., Tsai, C.-J. and Chiang, V.L. (1999) Repression of lignin biosynthesis promotes cellulose accumulation and growth in transgenic trees. *Nature Biotechnol.* 17, 808-812

Hua, Q., Araki, M., Koide, Y. and Shimizu, K. (2001) Effects of glucose, vitamins, and DO concentrations on pyruvate fermentation using *Torulopsis glabrata* IFO 0005 with metabolic flux analysis. *Biotechnol. Prog.* 17, 62-68.

Hunt, J.C. and Phibbs, P.V. Jr. (1983) Regulation of alternate peripheral pathways of glucose catabolism during aerobic and anaerobic growth of *Pseudomonas aeruginosa. J. Bacteriol.* 154, 793-802.

Iglesias, A.A., Plaxton, W.C. and Podestá, F.E. (1993) The role of inorganic phosphate in the regulation of C_4 photosynthesis. *Photosynth. Res.* 35, 205-211.

Iglesias, A.A. and Podestá, F.E. (1996) Photosynthate formation in crop plants. In *Handbook of Photosynthesis* (Pessarakli, M. ed.) pp. 681-698. Marcel Dekker Inc., New York.

Iglesias, A.A., Podestá, F.E. and Andreo C.S. (1997) Structural and regulatory properties of the enzymes involved in C_3, C_4, and CAM pathways for photosynthetic carbon assimilation. In: Handbook of Photosynthesis (Pessarakli, M. ed.), pp. 481-503. Marcel Dekker Inc., New York.

Ingraham, J.L., Maaloe, O. and Neidhardt, F.C. (1983) Chemical synthesis of the bacterial cell: Polymerization, biosynthesis, fuelling reactions, and transport, pp. 87-174. In *Growth of the Bacterial Cell.* Sinauer Associates, Sunderland, MA.

Ishizaki-Nishizawa, O., Fujii, T., Azuma, M., Sekiguchi, K. Murata, N. Ohtani, T. and Toguri, T. (1996) Low-temperature resistance of higher plants is significantly enhanced by a nonspecific cyanobacterial desaturase. *Nature Biotechnol.* 14, 1003-1006.

Jarman, T.R. and Pace, G.W. (1984) Energy requirements for microbial exopolysaccharide synthesis. *Arch. Microbiol.* 137, 231-235.

Jensen, K.F. and Pedersen, S. (1990) Metabolic growth rate control in *Escherichia coli* may be a consequence of subsaturation of the macromolecular biosynthetic apparatus with substrates and catalytic components. *Microbiol. Rev.* 54, 89-100.

Johnston, L.H. (1992) Cell cycle control of gene expression in yeast. *Trends Cell Biol.* 2, 353-357.

Jorgensen, H., Nielsen, J., Villadsen, J. and Mollgaard, H. (1995) Analysis of penicillin V biosynthesis during fed-batch cultivations with a high-yielding strain of *Penicillium chrysogenum. Appl. Microbiol. Biotechnol.* 43, 123-130.

Kacser, H. and J.A. Burns (1973) The control of flux. *Symp. Soc. Exp. Biol.*, 27, 65-104.

Kacser, H. and J.A. Burns (1981) The molecular basis of dominance. *Genetics*, 97, 639-

666.

Kacser, H. and J.W. Porteus (1987) Control of metabolism: What do we have to measure? *TIBS*, 12, 5-14.

Kaeppler, H.F., Gu, W., Somers, D.A., Rines, H.W. and Cockburn, A.F. (1990) Silicon carbide fiber-mediated DNA delivery into plant cells. *Plant Cell Rep.* 9, 415-418.

Kajiwara, S., Fraser, P.D., Kondo, K. and Misawa, N. (1997) Expression of an exogenous isopentenyl diphosphate isomerase gene enhances isoprenoid biosynthesis in *Escherichia coli. Biochem. J.* 324, 421-426.

Karp, P.D. (1998) Metabolic databases. *Trends Biochem. Sci.* 23, 114-116.

Kassir, Y., Granot, D. and Simchen, G. (1988) *IME1*, a positive regulator gene of meiosis in *S. cerevisiae. Cell* 52, 853-862

Kauffman, S.A. (1989) Adaptation on rugged fitness landscapes. In *Lectures in the Sciences of Complexity*, Vol. 1 (Stein, D.L., ed.). Addison-Wesley Publishing Co., Santa Fe Institute, New Mexico, pp. 527-712.

Kauffman, S. (1995) *At home in the Universe*. The search for the laws of self-organization and complexity. Oxford University Press, New York.

Keasling, J.D. (1999) Gene-expression tools for the metabolic engineering of bacteria. *TIBTECH* 17, 452-460.

Kell, D.B. and H.V. Westerhoff (1986) Metabolic control theory: its role in microbiology and biotechnology. *FEMS Microbiol. Rev.* 39, 305-320.

Kell, D.B., van Dam, K. and Westerhoff, H.V. (1989) Control analysis of microbial growth and productivity. In *Soc. Gen. Microbiol. Symp.*, pp. 44-93. Cambridge Univ. Press, London.

Keulers, M., Satroutdinov, A.D., Suzuki, T. and Kuriyama, H. (1996a) Synchronization effector of autonomous short-period-sustained oscillation of *Saccharomyces cerevisiae. Yeast* 12, 673-682.

Keulers, M., Suzuki, T., Satroutdinov, A.D. and Kuriyama, H. (1996b) Autonomous metabolic oscillation in continuous culture of *Saccharomyces cerevisiae* grown on ethanol. *FEMS Microbiol. Lett.* 142, 253-258.

Khetan, A., Malmberg, L.H., Sherman, D.H. and Hu, W.S. (1996) Metabolic engineering of cephalosporin biosynthesis in *Streptomyces clavuligerus. Ann. N. Y. Acad. Sci.* 782, 17-24.

Khetan, A. and Hu, W.-S. (1999) Metabolic engineering of antibiotic biosynthesis for process improvement. In Metabolic Engineering (Lee, S.Y. and Papoutsakis, E.T., eds.), pp. 177-202. Marcel Dekker, New York.

Kirschner, M. and Mitchinson, T. (1986) Beyond self-assembly: from microtubules to morphogenesis. *Cell* 45: 329-342.

Koizumi, S., Endo. T., Tabata, K. and Ozaki, A. (1998) Large-scale production of UDP-galactose and globotriose by coupling metabolically engineered bacteria. *Nature Biotechnol.* 16, 847-850.

Koshland Jr, D.E., Goldbeter, A. and Stock, J.B. (1982) Amplification and adaptation in regulatory and sensory systems. *Science* 217, 220-225.

Koshland, Jr., D.E. (1998) The era of pathway quantification. *Science* 280, 852-853.

Krens, F.A., Molendijk, L., Wullems, G.J. and Schilperoort, R.A. (1982) *In vitro* transformation of plant protoplasts with Ti-plasmid DNA. *Nature* 296, 72-74.

Ku, M.S.B., Agarie, S., Nomura, M., Fukuyama, H., Tsuchida, H., Ono, K., Hirose, S., Toki, S., Miyao, M. and Matsuoka, M. (1999) High-level expression of maize phosphoenolpyruvate carboxylase in transgenic rice plants, 17, 76-81.

Kuenzi, M. and Fiechter, A. (1969) Changes in carbohydrate composition and trehalase activity during the budding cycle of *Saccharomyces cerevisiae. Arch. Microbiol.* 64, 396-407.

Lamprecht, I. (1980) Growth and metabolism in yeasts. In *Biological Microcalorimetry.* (Beezer, A.E., ed.), pp. 43-112. Academic Press, London.

Lange, C.C., Wackett, L.P., Minton, K.W. and Daly, M.J. (1998) Engineering a recombinant *Deinococcus radiodurans* for organopollutant degradation in radioactive mixes waste environments. *Nature Biotechnol.* 16, 929-933.

Larsson, C., von Stockar, U., Marisson, I. and Gustafsson, L. (1993) Growth and metabolism of *Saccharomyces cerevisiae* in chemostat cultures under carbon-, nitrogen- or carbon- and nitrogen-limiting conditions. *J. Bacteriol.* 175, 4809-4816.

Lee, S.B. and Bailey, J.E. (1984) A mathematical model for sdv plasmid replication: Analysis of copy number mutants. *Plasmid* 11, 166-177.

Lee, J.Y., Roh, J.R. and Kim, H.S. (1994) Metabolic engineering of *Pseudomonas putida* for the simultaneous biodegradation of benzene, toluene, and *p*-xylene mixture. *Biotech. Bioengin.* 43, 1146-1152.

Lee, S.Y. and Papoutsakis, E.T. (1999) The challenges and promise of metabolic engineering. In *Metabolic Engineering* (Lee, S.Y. and Papoutsakis, E.T. eds.), pp. 1-12. Marcel Dekker, New York.

Lee, S.Y. and Choi, J.-I. (1999) Metabolic engineering strategies for the production of polyhydroxyalkanoates, a family of biodegradable polymers. In *Metabolic Engineering* (Lee, S.Y. and Papoutsakis, E.T. eds.), pp. 133-151. Marcel Dekker, New York.

Lessie, J.G. and Phibbs, Jr., P.V. (1984) Alternative pathways of carbohydrate utilization in pseudomonads. *Annu. Rev. Microbiol.* 38, 359-387.

Lew, D.J., Marini, N.J. and Reed, S.I. (1992) Different G1 cyclins control the timing of cell cycle commitment in mother and daughter cells of the budding yeast *S. cerevisiae. Cell* 69, 317-327.

Liao, J.C. and Delgado, J. (1993) Advances in metabolic control analysis. *Biotechnol. Prog.* 9, 221-233.

Liao, J.C., Hou, S.Y. and Chao, Y.P. (1996) Pathway analysis, engineering, and physiological considerations for redirecting central metabolism. *Biotech. Bioeng.* 52, 129-140.

Linton, J.D. and Stephenson, R.J. (1978) A preliminary study on growth yields in relation to the carbon and energy content of various organic growth substrates. *FEMS Microbiol. Lett.*, 3, 95-98.

Lloyd, D., Aon, M.A. and Cortassa S. (2001) Why homeodynamics, Not homeostasis? *TheScientificWorld* 1, 133-145.

Lloyd, D. and Gilbert, D.A. (1998) Temporal organization of the cell division cycle in eukaryotic microbes. In *Microbial Responses to Light and Time.* (Caddick, M.X., Baumberg, S., Hodgson, D.A. Phillips-Jones, M.K., eds.). Society for General Microbiology Symposium 56, Cambridge Univ. Press, UK.

Lloyd, D. (1998) Circadian and ultradian clock-controlled rhythms in unicellular microorganisms. *Adv. Microb. Physiol.* 39, 291-338.

Lloyd, A.L. and Lloyd, D. (1995) Chaos: its significance and detection in biology. *Biol. Rhythms Res.* 26, 233-252.

Lloyd, D. and Rossi, E.R. (1993) Biological rhythms as organization and information. *Biol. Rev.* 68, 563-577.

Lloyd, D. (1992) Intracellular Time Keeping: Epigenetic Oscillations reveal the functions of an Ultradian Clock. In *Ultradian Rhythms in Life Processes* (Lloyd, D., Rossi, E.R., eds), pp. 5-22. Springer-Verlag, London.

Lloyd, D., Poole, R.K. and Edwards, S.W. (1982) *The Cell Division Cycle: Temporal Organisation and Control of Cellular Growth and Reproduction.* Academic Press, London.

Lloyd, D. (1974) *The Mitochondria of Microorganisms*, pp. 160-167. Academic Press, London.

Lloyd, J.R., Springer, F., Buléon, A., Müller-Rober, B., Willmitzer, L. and Kossmann, J. (1999) The influence of alterations in ADPglucose pyrophosphorylase activities on starch structure and composition in potato tubers. *Planta* 209, 230-238.

Lopez de Felipe, F., Kleerebezem, M., de Vos, W.M. and Hugenholtz, J. (1998) Cofactor engineering: a novel approach to metabolic engineering in *Lactococcus lactis* by controlled expression of NADH oxidase. *J. Bacteriol.* 180, 3804-3808.

Lundblad, R.L. and Kingdon, H.S. (1999) Molecular pharming. *Biotechnol. Appl. Biochem.* 30, 99-100.

MacFarlane, A.G.J. (1973) *Dynamical Systems Models*. Harrap, London.

Mark, A., de Graaf, A.A., Wiechert, W., Eggeling, L. and Sahm, H. (1996) Determination of the fluxes in the central metabolism of *Corynebacterium glutamicum* by nuclear magnetic resonance spectroscopy combined with metabolite balancing. *Biotech. Bioengin.* 49, 111-129.

Martegani, E., Porro, D., Ranzi, B.M. and Alberghina, L. (1990) Involvement of cell size control mechanisms in the induction and maintenance of oscillations in continuous cultures of yeast. *Biotech. Bioengin.* 36, 453-459.

McGaughey, W.M. and Whalon, M.E. (1992) Managing insect resistant to *Bacillus thuringiensis* toxins. *Science* 258, 1451-1455.

Miklos, G.L.G. and Rubin, G.M. (1996) The role of the Genome Project in determining gene function: Insights from model organisms. *Cell* 86, 521-529.

Miller, J. J. (1989) Sporulation in *Saccharomyces cerevisiae*. In *The Yeast*, 2nd ed, vol 2 (Rose, A.H. and Harrison, J.S., eds.), pp. 489-550. Academic Press, London.

Mitchison, T. and Kirschner, M. (1984) Dynamic instability of microtubules growth. *Nature* 312, 237-242.

Mittendorf, V., Robertson, E.J., Leech,R.M., Krüger, N., Steinbüchel, A. and Poirier Y. (1998) Synthesis of medium-chain-length polyhydroxyalkanoates in *Arabidopsis thaliana* using intermediates of peroxisomal fatty acid β-oxidation. *Proc. Natl. Acad. Sci. USA* 95, 13397-13402.

Mol, J.N.M., Holton, T.A. and Koes, R.E. (1995) Floriculture: genetic engineering of commercial traits. *Trends Biotechnol.* 13, 350-355.

Mónaco, M.E., Valdecantos, P. and Aon, M.A. (1995) Carbon and energetic uncoupling associated with cell cycle arrest of cdc mutants of *Saccharomyces cerevisiae* may be linked to glucose-induced catabolite repression. *Exp. Cell Res.* 217, 52-56.

Mónaco, M.E. (1996) Crecimiento y proliferación de *Saccharomyces cerevisiae*. PhD Thesis, Universidad Nacional de Tucumán, Argentina.

Monod, J. (1950) La technique de culture continue: theorie et applications. *Ann. Inst. Pasteur* 79, 390-410.

Morgan, J.A., Rijhwani, S.K. and Shanks, J.V. (1999) Metabolic engineering for the production of plant secondary metabolites. In Metabolic Engineering (Lee, S.Y. and Papoutsakis, E.T., eds.), pp. 325-351. Marcel Dekker, New York.

Münch, T., Sonnleitner, B. and Fiechter, A. (1992) New insights into the synchronization mechanism with forced synchronous cultures of *Saccharomyces cerevisiae*. *J. Biotechnol.* 24, 299-314.

Murai, T., Ueda, M., Yamamura, M., Atomi, H., Shibasaki, Y., Kamasawa, N., Osumi, M., Amachi, T. and Tanaka, A. (1997) Construction of a starch-utilizing yeast by cell surface engineering. *Appl. Environm. Microbiol.* 63, 1362-1366.

Murray, D.B., Engelen, F.A., Keulers, M., Kuriyama, H. and Lloyd, D. (1998) NO$^+$, but not NO, inhibits respiratory oscillations in ethanol-grown chemostat cultures of *Saccharomyces cerevisiae*. *FEBS Lett.* 431, 297-299.

Murray, D.B., Engelen, F.A., Lloyd, D. and Kuriyama, H. (1999) Involvement of glutathione in the regulation of respiratory oscillations during a continuous culture of *Saccharomyces cerevisiae*. *Microbiology* 145, 2739-2745.

Mushegian, A.R. and Koonin, E.V. (1996) A minimal gene set for cellular life derived by comparison of complete bacterial genomes. *Proc. Natl. Acad. Sci. USA* 93, 10268-10273.

Neuhaus, G. and Spangenberg, G. (1990) Plant transformation by microinjection techniques. *Physiol. Plant.* 79, 213-217.

Newcomb, W. (1990) Plastid structure and development. In *Plant Physiology, Biochemistry and Molecular Biology* (Dennis, D.T.; Turpin, D.H., eds.). Chap. 13, pp. 193-197. Longman Scientific & Technical, England.

Nicolis, G. and Prigogine, I. (1977) *Self-organization in Non-equilibrium Systems*. Wiley-Interscience, London

Nicolis, G. and Prigogine, I. (1989) *Exploring Complexity*. Freeman, New York.

Niederberger, P., Prasad, R., Miozzari, G. and Kacser, H. (1992) A strategy for increasing an *in vivo* flux by genetic manipulation. *Biochem. J.* 287, 473-479.

Novick, A. and Szilard, I. (1950) Description of the chemostat. *Science* 112, 715-716-

Nuccio, M.L., Rhodes, D., McNeil, S.D. and Hanson, A.D. (1999) Metabolic engineering of plants for osmotic stress resistance. *Curr. Opinion Plant Biol.* 2, 128-134.

Nyiri, L.K. (1972) Applications of computers in biochemical engineering. In *Adv. Biochem. Engin.* Springer Verlag, New York.

Ohlrogge, J. (1999) Plant Metabolic Engineering: are we ready for phase two? *Current Op. Plant Biol.* 2, 121-122.

Oliver, S.G. (1996) From DNA sequence to biological function. *Nature* 379, 597-600.

Oliver, S.G. (2000) Guilt-by-association goes global. *Nature* 403, 601-603.

Overbeek, R., Larsen, N., Pusch, G.D., D'Souza, M., Selkov, E. Jr., Kyrpides, N., Fonstein, M., Maltsev, N. and Selkov, E. (2000) WIT: integrated system for high-throughput genome sequence analysis and metabolic reconstruction. *Nucleic Ac. Res.* 28, 123-125.

Ott, E., Sauer, T. and Yorke, J.A. (1994) *Coping with Chaos*. Wiley and Sane, New York.

Ovadi, J. And Srere, P.A. (2000) Macromolecular compartmentation and channeling. *Int. Rev. Cytol.* 192, 255-280.

Owen, M.R.L. and Pen, J., Eds. (1996) *Transgenic Plants: A Production System for Industrial and Pharmaceutical Proteins*, 348 pp., J. Wiley & Sons, New York.

Ozcan, S. and Johnston, M. (1999) Function and regulation of yeast hexose transporters. *Microbiol. Mol. Biol. Rev.* 63, 554-569.

Pagé N., Kluepfel, D., Shareck, F. and Morosoli, R. (1996) Increased xylanase yield in *Streptomyces lividans*: Dependence on number of ribosome-binding sites. *Nature Biotechnol.* 14, 756-759.

Papoutsakis, E.T. and Bennett, G.N. (1999) Molecular regulation and metabolic engineering of solvent production by *Clostridium acetobutylicum*. In *Metabolic Engineering* (Lee, S.Y. and Papoutsakis, E.T., eds.), pp. 253-279. Marcel Dekker, New York.

Park, S.J., McCabe, J., Turna, J. and Gunsalus, R.P. (1994) Regulation of the citrate synthase (*gltA*) gene of *Escherichia coli* in response to anaerobiosis and carbon supply: Role of the *arcA* gene product. *J. Bacteriol.* 176, 5086-5092.

Park, S.J., Cotter, P.A. and Gunsalus, R.P. (1995) Regulation of malate dehydrogenase (*mdh*) gene expression in *Escherichia coli* in response to oxygen, carbon, and heme availability. *J. Bacteriol.* 177, 6652-6656.

Park, S.M., Sinskey, A.J. and Stephanopoulos, G. (1997) Metabolic and physiological studies of *Corynebacterium glutamicum* mutants. *Biotech. Bioengin.* 55, 864-879.

Parulekar, S.J., Semones, G.B., Rolf, M.L., Lievense, J.C. and Lim, H.C. (1986) Induction and elimination of oscillations in continuous cultures of *Saccharomyces cerevisiae*. *Biotech. Bioengin.* 28, 700-710.

Peng, B., Petrov, V. and Showalter, K. (1991) Controling chemical chaos. *J. Phys. Chem.* 95, 4957-4959.

Penman, S., Fulton, A., Capco, D. *et al.* (1981) Cytoplasmic and nuclear architecture in cells and tissue: forms, functions, and mode of assembly. *Cold Spring Harbor Symp. Quant. Biol.* 45, 1013-1028.

Pierik, R.L.M. (1988) In Vitro *Culture of Higher Plants*, 344 pp. Kluwer Academic Publishers, Dordrecht.

Pines, O., Shemesh, S., Battat, E. and Goldberg, I. (1997) Overexpression of cytosolic malate dehydrogenase (MDH2) causes overproduction of specific organic acids in *Saccharomyces cerevisiae. Appl. Microbiol. Biotechnol.* 48, 248-255.

Pirt, S.J. (1975) *Principles of Microbe and Cell Cultivation*. Blackwell Scientific Public., London.

Plaxton, W.C. (1996) The organization and regulation of plant glycolysis. *Annu. Rev. Plant Physiol. Plant Mol. Biol.* 47, 185-214.

Platteeuw, Ch., Hugenholtz, J., Starrenburg, M., Van Alen-Boerrigter, I. and De Vos, W.M. (1995) Metabolic engineering of *Lactococcus lactis*: Influence of the overproduction of α-acetolactate synthase in strains deficient in lactate dehydrogenase as a function of culture conditions. *Appl. Environm. Microbiol.* 61, 3967-3971.

Poirier, Y. (1999) Production of new polymeric compounds in plants. *Curr. Opin. Biotechnol.* 10,181-185.

Porro, D., Martegani, E., Ranzi, B.M. and Alberghina, L. (1988) Oscillations in continuous cultures of budding yeast: A segregated parameter analysis. *Biotech. Bioengin.* 32, 411-417.

Porro, D., Brambilla, L., Ranzi, B.M., Martegani, E. and Alberghina, L. (1995) Development of metabolically engineered *Saccharomyces cerevisiae* cells for the production of lactic acid. *Biotechnol. Prog.* 11, 294-298.

Raamsdonk, L.M., Teusink, B., Broadhurst, D., *et al.* (2001) A functional genomics strategy that uses metabolome data to reveal the phenotype of silent mutations. *Nature Biotechnol.* 19, 45-50.

Ramos, J.L., Díaz, E., Dowling, D., de Lorenzo, V., Molin, S., O'Gara, F., Ramos, C. and Timmis, K.N. (1994) The behavior of bacteria for biodegradation. *Bio/Technology* 12, 1349-1356.

Reddy, A.S. and Thomas, T.L. (1996) Expression of a cyanobacterial delta6 desaturase gene results in gamma-linolenic acid production in transgenic plants. *Nature Biotechnol.* 14, 639-642.

Reed, S.I. (1992) The role of p34 kinases in the G1 to S-phase transition. *Annu. Rev. Cell Biol.* 8, 529-561.

Reich, J.G. and Sel'Kov, E.E. (1981) *Energy Metabolism of the Cell*. Academic Press. London.

Richardson, H., Lew, D.J., Henze, M., Sugimoto, K. and Reed, S.I. (1992) Cyclin-B homologs in *Saccharomyces cerevisiae* function in S phase and in G2. *Genes & Dev.* 6, 2021-2034.

Robson, P.R.H., McCormac, A.C., Irvine, A.S. and Smith, H. (1996) Genetic engineering of harvest index in tobacco through overexpression of a phytochrome gene. *Nature Biotechnol.* 14, 995-998.

Roels, J.A. (1983) *Energetics and Kinetics in Biotechnology*. Elsevier, Amsterdam.

Ronne, H. (1995) Glucose repression in fungi. *Trends Genet.* 11, 12-17.

Rosen, R. (1967) *Optimality Principles in Biology*. Butterworths, London.

Rosen, R. (1970) *Dynamical System Theory in Biology*. John Wiley & Sons, New York.

Rouwenhorst, R.J., Visser, L.E., van der Baan, A.A., Scheffers, W.A. and van Dijken, J.P. (1988) Production, distribution and kinetic properties of inulinase in continuous cultures of *Kluyveromyces marxianus* CBS 6556. *Appl. Environm. Microbiol.* 54, 1131-1137.

Rugh,C.L., Senecoff, J.F., Meagher, R.B. and Merkle, S.A. (1998) Development of a transgenic yellow poplar for mercury phytoremediation. *Nature Biotechnol.* 16, 925-928.

Saks, V.A., Kaambre, T., Sikk, P. *et al.* (2001) Intracellular energetic units in red muscle cells. *Biochem. J.* 356, 643-657.

Satroutdinov, A.D., Kuriyama, H. and Kobayashi, H. (1992) Oscillatory metabolism of *Saccharomyces cerevisiae* in continuous culture. *FEMS Microbiol. Lett.* 77, 261-267.

Sauer, U., Hatzimanikatis, V., Bailey, J.E., Hochuli, M., Szyperski, T. and Wülhrich, K. (1997) Metabolic fluxes in riboflavin-producing *Bacillus subtilis*. *Nature Biotechnol.* 15, 448-452.

Sauer, U., Cameron, D.C. and Bailey, J.E. (1998) Metabolic capacity of *Bacillus subtilis* for the production of purine nucleosides, riboflavin, and folic acid. *Biotech. Bioengin.* 59, 227-238.

Sauro, H.M., Small, J.R. and Fell, D.A. (1987) Metabolic control and its analysis. Extensions to the theory and matrix method. *Eur. J. Biochem.*, 165, 215-221.

Savageau, M.A. (1976) *Biochemical Systems Analysis: A Study of Function and Design in Molecular Biology*. Reading MA, Addison Wesley.

Savageau, M.A. (1991) Biochemical Systems Theory: operational differences among variant representations and their significance. *J. Theor. Biol.* 151, 509-530.

Savinell, J.M. and Palsson, B.O. (1992a) Optimal selection of metabolic fluxes for *in vivo* measurement. I. Development of mathematical methods. *J. Theor. Biol.* 155, 201-214.

Savinell, J.M. and Palsson, B.O. (1992b) Network analysis of intermediates using linear optimisation. I. Development of mathematical formalism. *J. Theor. Biol.* 154, 421-454.

Savinell, J.M. and Palsson, B.O. (1992c) Network analysis of intermediates using linear optimisation. II. Interpretation of hybridoma cell metabolism. *J. Theor. Biol.* 154, 455-473.

Schüller, H.J. and Entian, K.D. (1987) Isolation and expression analysis of two yeast regulatory genes involved in the derepression of glucose-repressible enzymes. *Mol. Gen. Genet.* 209, 366-373.

Schüller, H.J. and Entian, K.D. (1991) Extragenic suppressors of yeast glucose derepression mutants leading to constitutive synthesis of several glucose-repressible enzymes. *J. Bacteriol.* 173, 2045-2052.

Schuster, H.G. (1988) *Deterministic Chaos: An Introduction*. VCH, Zurich.

Sel'kov, E.E. (1968) Self-oscillations in glycolysis. A simple kinetic model. *Eur. J. Biochem.* 4, 79-86.

Senior, P.J., Beech, G.A., Ritchie, G.A.F. and Dawes, E.A. (1972) The role of oxygen limitation in the formation of poly-β-hydroxybutyrate during batch and continuous culture of *Azotobacter beijerinckii*. *Biochem. J.* 128, 1193-1201.

Sevenier, R., Hall, R.D., van der Meer, I.M. ,Hakkert, H.J.C., van Tunen, A.J. and Koops, A.J (1998) High level fructan accumulation in a transgenic sugar beet. *Nature Biotechnol.* 16, 843-846.

Shi, N.Q. and Jeffries, T.W. (1998) Anaerobic growth and improved fermentation of *Pichia stipitis* bearing a URA1 gene from *Saccharomyces cerevisiae*. *Appl.*

Microbiol. Biotechnol. 50, 339-345.

Shimada, H., Kondo, K., Fraser, P.D., Miura, Y., Saito, T. and Misawa, N. (1998) Increased carotenoid production by the food yeast *Candida utilis* through metabolic engineering of the isoprenoid pathway. *Appl. Environm Microbiol.* 64, 2676-80.

Shinbrot, T., Grebogi, C., Ott, E. and Yorke, J.A. (1993) Using small perturbations to control chaos. *Nature* 363, 411-417.

Shintani, D. and Della Penna, D. (1998) Elevating the vitamin E content of plants through metabolic engineering. *Science* 282, 2098-2100.

Shu, J. and Schuler, M.L. (1989) A mathematical model for the growth of a single cell of *E. coli* on a glucose/glutamine/ammonium medium. *Biotech. Bioengin.* 37, 1117-1126.

Shu, J. and Schuler, M.L. (1991) Prediction of the effects of amino acid supplementation on growth of *E. coli*. B/r. *Biotech. Bioengin.* 37, 708-715.

Sierkstra, L.N., Verbakel, J.M.A. and Verrips, C.T. (1992) Analysis of transcription and translation of glycolytic enzymes in glucose-limited continuous cultures of *Saccharomyces cerevisiae*. *J. Gen. Microbiol.* 138, 2559-2566.

Sivak, M.N., and Preiss, J. (1998) Advances in Food Nutrition Research. Vol. 41, *Starch: Basic Science to Biotechnology*, 199 pp. Academic Press, San Diego.

Slater, J.H. (1985) Microbial growth dynamics. In *Comprehensive Biotechnology*. The principles, applications and regulations of biotechnology in industry, agriculture and medicine (Moo-Young, M., Bull, A.T. and Dalton, H., eds.). Vol. 1, pp. 189-213. Pergamon Press, Oxford.

Smirnoff, N. (1998) Plant resistance to environmental stress. *Curr. Opin. Biotechnol.* 9, 214-219.

Smits H.P., Hauf, J., Muller, S., Hobley, T.J., Zimmermann, F.K., Hahn-Hagerdal, B., Nielsen, J. and Olsson, L. (2000) Simultaneous overexpression of enzymes of the lower part of glycolysis can enhance the fermentative capacity of *Saccharomyces cerevisiae*. *Yeast*, 16. 1325-1334

Sode, K., Sugimoto, S., Watanabe, M. and Tsugawa, W. (1995) Effect of PQQ glucose dehydrogenase overexpression in *Escherichia coli* on sugar-dependent respiration. *J. Biotechnol.* 43, 41-44.

Somerville, C.R. and Bonetta, D. (2001) Plants as factories for technical materials. *Plant Physiol.* 125, 168-171.

Somerville, C.R. and Briscoe, J. (2001) Genetic engineering and water. *Science* 292, 2217.

Sonnewald U., Hajirezaei, M-R., Kossmann, J., Heyer A., Trethewey, R.N. and Willmitzer L. (1997) Increased potato tuber size resulting from apoplastic expression of a yeast invertase. *Nature Biotechnol.* 15, 794-797.

Sonnewald, U., and Willmitzer, L. (1992) Molecular approaches to sink-source interactions. *Plant Physiol.* 99, 1267-1270.

Srere, P.A. (1993) Wanderings (Wonderings) in metabolism. *Biol. Chem. Hoppe-Seyler* 374, 833-842.

Standing, C.N., Fredrickson, A.G. and Tsuchiya, H.M. (1972) Batch- and continuous-culture transients for two substrate systems. *App. Microbiol.* 23, 354-359.

Stark, D.M., Timmerman, K.P., Barry, G.F., Preiss, J. and Kishore, G.M. (1992) Regulation of the amount of starch in plant tissues by ADPGlucose Pyrophosphorylase. *Science* 258, 287-292.

Staskawicz, B.J., Ausubel, F.M., Baker, B.J., Ellis, J.G. and Jones J.D.G. (1995) Molecular genetics of plant disease resistance. *Science* 268, 661-667.

Stephanopoulos, G.N. and Vallino, J.J. (1991) Network rigidity and metabolic engineering in metabolite overproduction. *Science*, 252, 1675-1681.

Stephanopoulos, G.N., Aristidou, A.A. and Nielsen, J. (1998) *Metabolic Engineering. Principles and Methodologies.* Academic Press, San Diego.

Stephanopoulos, G.N. (2000) Bioinformatics and metabolic engineering. *Metab. Engin.* 2, 157-158.

Stitt, M. (1994) Manipulation of carbohydrate partitioning. *Curr. Opin. Biotechnol.* 5, 137-143.

Stitt, M. (1999) The first will be the last and the last will be first: Non-regulated enzymes call the tune? In *Plant Carbohydrate Biochemistry* (Bryant, J.A., Burnell, M.M., and Kruger, N.J., eds.). Chap. 1, pp. 1-16. BIOS Scientific Publishers Ltd., Oxford.

Stoop, J.M.H., Williamson, J.D. and Pharr, D.M. (1996) Mannitol metabolism in plants: A method for coping with stress. *Trends Plant Sci.* 1, 139-144.

Stouthamer, A.H. (1979) The search for correlation between theoretical and experimental growth yields. *Int. Rev. Biochem.*, 21, 1-47.

Stouthamer, A.H. and van Verseveld, H.W. (1985*) Stoichiometry of microbial growth. In Comprehensive Biotechnology.* The principles, applications and regulations of biotechnology in industry, agriculture and medicine (Moo-Young, M., Bull ,A.T and Dalton, H., eds.). Vol. 1, pp. 215-238. Pergamon Press, Oxford.

Stouthamer, A.H. and van Verseveld, H.W. (1987) Microbial energetics should be considered in manipulating metabolism for biotechnological purposes. *TIBTECH*, 5, 149-155.

Strassle, C., Sonnleitner, B. and Fiechter, A. (1988/89) A predictive model for the spontaneous synchronization of *Saccharomyces cerevisiae. J. Biotechnol.* 7, 299-318.

Strom, A.R. (1998) Osmoregulation in the model organism *Escherichia coli*: genes governing the synthesis of glycine betaine and trehalose and their use in metabolic engineering of stress tolerance. *J. Biosci.* 23, 437-445

Surana, U., Robitsch, H., Price, C., Schuster, T., Fitch, I., Futcher, A.B. and Nasmyth, K. (1991) The role of *CDC28* and cyclins during mitosis in the budding yeast *S. cerevisiae. Cell* 65, 145-161.

Tempest, D.W. (1965) Magnesium-limited growth of *Aerobacter aerogenes* in a chemostat. *J. Gen. Microbiol.* 39, 355-366.

Tempest, D.W. and Neijssel, O.M. (1984) The status of Y_{ATP} and maintenance energy as biologically interpretable phenomena. *Annu. Rev. Microbiol.* 38, 459-486.

Tepfer, D. (1990) Genetic transformation using *Agrobacterium rhizogenes. Physiol. Plant.* 79, 140-146.

Theologis, A. (1994) Control of ripening. *Curr. Opin. Biotechnol.* 5, 152-157.

Topfer, R., Martini, N. and Schell, J. (1995) Modification of plant lipid synthesis. *Science* 268, 681-686.

Toivari, M.H., Aristidou, A., Rouhonen, L. and Penttila, M. (2000) Conversion of xylose to ethanol by recombinant *Saccharomyces cerevisiae*: Importance of xylulokinase (XKS1) and oxygen availability. *Metab. Engin.* 3, 236-249.

Topiwala, H.H. and Hamer, G. (1971) Effects of wall-growth in steady state continuous culture. *Biotech. Bioengin.* 13, 919-922.

Treinin, M. and Simchen, G. (1993) Mitochondrial activity is required for the expression of *IME1*, a regulator of meiosis in yeast. *Curr. Genet.* 23, 223-227.

Tsuchiya, H.M. (1983) The holding time in pure and mixed culture fermentations. *Ann. New York Acad. Sci.* 413, 184-192.

Vallino, J.J. and Stephanopoulos, G. (1993) Metabolic flux distributions in *Corynebacterium glutamicum* during growth and lysine overproduction. *Biotech. Bioeng.* 41, 633.646.

Van Dam, K. (1996) Role of glucose signalling in yeast metabolism. *Biotech. Bioeng.* 52, 161-165.

Van Dien, S.J. and Keasling, J.D. (1998) Optimisation of polyphosphate degradation and phosphate secretion using hybrid metabolic pathways and engineered host strains. *Biotech. Bioengin.* 59, 754-761.

Van Dijken, J.P., Otto, R. and Harder, W. (1976) Growth of *Hansenula polymorpha* in a methanol limited chemostat: physiological responses due to the involvement of methanol oxidase as a key enzyme in methanol metabolism. *Arch. Microbiol.* 111, 137-144.

Vanrolleghem, P.A., Jong-Gubbels, P. van Gulik, W.M., Pronk, J.T. van Dijken, J.P. and Heijnen, S. (1996) Validation of a metabolic network for *Saccharomyces cerevisiae* using mixed substrate studies. *Biotechnol. Prog.* 12, 434-448.

Varma, A., Boesch, B.W. and Palsson, B.O. (1993) Stoichiometric interpretation of *Escherichia coli* glucose catabolism under various oxygenation rates. *Appl. Environm. Microbiol.* 59, 2465-2473.

Varma, A. and Palsson, B.O. (1994) Metabolic Flux Balancing: Basic concepts, scientific and practical use. *Biotechnology*, 12, 994-998.

Varner, J. and Ramkrishna, D. (1999) Metabolic engineering from a cybernetic perspective. 1. Theoretical preliminaries. *Biotechnol. Prog.* 15, 407-425.

Vaseghi, S., Baumeister, A., Rizzi, M. and Reuss, M. (1999) *In vivo* dynamics of the pentose phosphate pathway in *Saccharomyces cerevisiae*. *Metab. Engin.* 1, 128-140.

Verdoni, N., Aon, M.A., Lebeault, J.M. and Thomas, D. (1990) Proton motive force, energy recycling by end product excretion, and metabolic uncoupling during anaerobic growth of *Pseudomonas mendocina*. *J. Bacteriol.* 172, 6673-6681.

Verdoni, N., Aon, M.A. and Lebeault, J.M. (1992) Metabolic and energetic control of *Pseudomonas mendocina* growth during transitions from aerobic to oxygen-limited conditions in chemostat cultures. *Appl. Environm. Microbiol.* 58, 3150-3156.

Visser, R.G.F. and Jacobsen, E. (1993) Towards modifying plants for altered starch content and composition. *Trends Biotechnol.* 11, 63-68.

Von Klitzing, L. and Betz, A. (1970) Metabolic control in flow systems. 1. Sustained glycolytic oscillation in yeast suspension under continuous substrate infusion. *Arch. Mikrobiol.* 71, 220-225.

Von Meyenburg, K. (1973) Stable synchrony oscillations in continuous cultures of *Saccharomyces cerevisiae* under glucose limitations. In Biological and biochemical oscillators (Chance, B., Pye, E.K., Ghosh, A.K. and Hess, B., eds.), pp. 411-417. Academic Press, New York.

Vriezen, N. and van Dijken, J.P. (1998) Fluxes and enzyme activities in central metabolism of myeloma cells grown in chemostat culture. *Biotech. Bioeng.* 59, 28-39.

Walker, D.A. (1995) Manipulating photosynthesis metabolism to improve crops: An inversion of ends and means. *J. Exp. Bot.* 46, 1253-1259.

Walsh, M.C., Scholte, M., Valkier, J., Smits, H.P. and van Dam, K. (1996) Glucose sensing and signalling properties in *Saccharomyces cerevisiae* require the presence of at least two members of the glucose transporter family. *J. Bacteriol.* 178, 2593-2597.

Wang, C.L., Cooney, A.L., Demain, P., Dunnill, A., Ehumphrey, M. and Lilly, D. (1979) *Fermentation and Enzyme Technology*. John Wiley, New York.

Wang, X. and Da Silva, N.A. (1996) Site-specific integration of heterologous genes in yeast via Ty3 retrotransposition. *Biotech. Bioeng.* 51, 703-712.

Wei, M.L., Webster, D.A. and Stark, B.C. (1998) Metabolic engineering of *Serratia marcescens* with the bacterial hemoglobin gene: alterations in fermentation pathways. *Biotech. Bioeng.* 59, 640-646.

Weng, G., Bhalla, U.S. and Iyengar, R. (1999) Complexity in biological signalling systems. *Science* 284, 92-96.

Westerhoff, H.V., Lolkema, J.S., Otto, R. and Hellingwerf, K.J. (1982) Thermodynamics of growth. Non-equilibrium thermodynamics of bacterial growth. The phenomenological and the mosaic approach. *Biochim. Biophys. Acta*, 683, 181-220.

Westerhoff, H.V. and Kell, D.B. (1987) Matrix method for determining steps most rate-limiting to metabolic fluxes in biotechnological processes. *Biotech. Bioengin.* 30, 101-107.

Westerhoff, H.V. and van Dam, K. (1987) *Thermodynamics and Control of Biological Free-energyTtransduction*. Elsevier, Amsterdam.

Weusthuis, R.A., Pronk, J.T., van den Broek, P.J. and van Dijken, J.P. (1994) Chemostat cultivation as a tool for studies on sugar transport in yeasts. *Microbiol. Rev.* 58, 616-630.

Willmitzer, L. (1999) Plant Biotechnology: output traits the second generation of plant biotechnology products is gaining momentum. *Curr. Opin. Biotechnol.* 10, 161-162.

Wilson, L.P. and Bouwer, E.J. (1997) Biodegradation of aromatic compounds under mixed oxygen/denitrifying conditions: a review. *J. Ind. Microbiol & Biotechnol.* 18, 116-130.

Withers, J.M., Swift, R.J., Wiebe, M.G., Robson, G.D., Punt, P.J., Van den Hondel, C.A.M.J.J. and Trinci, A.P.J. (1998) Optimisation and stability of glucoamylase production by recombinant strains of *Aspergillus niger* in chemostat culture. *Biotech. Bioengin.* 59, 407-418.

Wodicka, L., Dong, H., Mittmann, M., Ho, M.-H. and Lockhart, D.J. (1997) Genome-wide expression monitoring in *Saccharomyces cerevisiae*. *Nature Biotechnol.* 15, 1359-1367.

Woods, A., Munday, M.R., Scott, J., Yang, X., Carlson, M. and Carling, D. (1994) Yeast *SNF1* is functionally related to mammalian AMP-activated protein kinase and regulates acetyl-CoA carboxylase *in vivo*. *J. Biol. Chem.* 269, 19509-19515.

Yarmush, M.L. and Berthiaume, F. (1997) Metabolic engineering and human disease. *Nature Biotechnol.* 15, 525-528.

Ye, X., Al-Babili, S. Klöti, A., Zhang, J., Lucca, P., Beyer, P. and Potrykos, I. (2000) Engineering the provitamin A (ß-carotene) biosynthetic pathway into (carotenoid-free) rice endosperm. *Science* 287, 303-305.

Zaslavskaia, L.A., Lippmeier, J.C., Shih, C., Ehrhardt, D., Grossman, A.R. and Apt, K.E. (2001) Throphic conversion of an obligate photoautotrophic organism through metabolic engineering. *Science* 292, 2073-2075.

Zhang, M., Eddy, Ch., Deanda, K., Finkelstein, M. and Picataggio, S. (1995) Metabolic engineering of a pentose metabolism pathway in ethanologenic *Zymomonas mobilis*. *Science* 267, 240-243.

Zimmermann, F.K., Kaufmann, I., Rasenberger, H. and Haussmann, P. (1977) Genetics of carbon catabolite repression in *Saccharomyces cerevisiae*: genes involved in the derepression process. *Mol. Gen. Genet.* 151, 95-103.

Zupan, J.R. and Zambryski, P. (1995) Transfer of T-DNA from *Agrobacterium* to the Plant Cell. *Plant Physiol.* 107, 1041-1047.

Index

α-acetolactate 124, 125, 126
α-acetolactate decarboxylase 124, 126
α-acetolactate synthase 124, 126
α-galactosidase 220
α ketoglutarate 103
Agrobacterium
    rhizogenes 184
    tumefaciens 177, 184
AcCoA 58, 59, 60, 61, 117, 206
AcCoA synthetase 117
acetaldehyde dehydrogenase 57
Acetobacter aceti 40
acetoin 8, 10, 11, 15, 16, 124, 126, 160
actual yields 7
adiabatic 65
ADPGlcPPase 134, 179, 180, 197, 198, 199, 200, 201
ADPglucose 179
ADPglucose pyrophosphorylase 179
    See also ADPGlcPPase
Aerobacter aerogenes 44, 72, 81
Agaricus 3
alanine 13, 56, 61
alcohol dehydrogenase 27, 121, 211
aldolase 191, 197, 211
alginate 84, 85, 208, 209
alkaloids 15, 180
allometric law 23
allosteric
    effectors 134, 198, 199, 201
    regulation 25
ambiquity 133
amino acids 10, 57, 58, 71, 78, 80, 97, 118, 138, 177, 181, 184
    aromatic 10, 107, 110, 205
amylopectin 180
amyloplasts 180, 186, 188, 198
amylose 180
antibiotics 2, 119
apoptosis 214
Arabidopsis 13, 20, 181
arabinose 7, 9, 13, 121, 122
aromatic compounds 11, 62

aspartate 61, 193
Aspergillus 3, 83
autocatalytic 30, 31, 32, 33, 94, 96, 191, 213
Azotobacter beijerincki 209

β-carotene 181
β-lactam antibiotics 10
Bacillus thuringiensis 177
bacteria 3, 8, 9, 24, 26, 27, 33, 46, 106, 116, 123, 156, 159, 167, 173, 178, 181, 184, 235
balance
    carbon 48
    differential mass 41
    dynamic 91, 93, 94
    electron 47, 48
    elemental 41, 45, 48, 55, 117
    integral 41
    mass 4, 56
bifurcation 19, 21, 22, 32, 34, 35, 36, 96, 135
    diagram 85, 135
    Hopf 21, 135
biochemical system theory 4
biological organization 17, 30
biomass
    concentration 85, 86, 90, 92, 99, 120, 138
    microbial 23, 44
    synthesis 42, 45, 47, 56, 72, 80, 97, 103, 111, 216
bistability 21, 33, 157
Black Box 39
butanediol 15, 126
butanol 9, 122
by-products 5, 10, 42, 48, 85, 102, 106, 119

Caenorhabditis 17, 20
Calorimetry 67
Calvin cycle 134
Candida utilis 12, 44, 73, 167
carbon dioxide 24, 168, 169

catabolite repression 10, 29, 138, 140, 141,
        205, 206, 213, 217, 218, 219,
        223
cdc mutants   216, 217, 218
CDC28   18, 209, 216, 217
    -lacZ fusion   219
cell
    bundle-sheath   193
    cycle arrest   95
    differentiation   10, 29
    division cycle 10, 18, 166, 209, 216, 218
    mesophyll   193
cellular engineering   1, 2, 42, 203, 214
cellular scaffolds   25, 29, 204
chaotic   19, 21, 37, 157
    attractor   171
    dynamics   37
checkpoints   219
Chlamydomonas reinhardtii   187
chlorophyll pigments   186
circadian rhythms   34
Clarkia xantiana   197
coherent   33, 34, 35
compartmentation   31, 179, 186, 188, 197
connectivity theorems   131, 132
control coefficients 7, 127, 128, 129, 130,
        131, 132, 133, 135, 140, 141,
        201, 210, 212, 213
    concentration-   129, 134, 212
    flux-   100, 128, 129, 131, 133, 140
Corynebacterium   3, 13, 115
coupling   21, 23, 29, 34, 35, 37, 47, 48, 93,
        95, 119, 122, 214, 215, 216
covalent modification   25, 28, 33
crassulacean acid metabolism   195
critical dilution rate   103, 104, 137
culture
    batch 4, 7, 12, 39, 40, 41, 69, 70, 77,
        78, 79, 80, 81, 82, 90, 91, 92, 98,
        111, 120, 121, 154, 158, 217, 219,
        220
    chemostat 5, 7, 20, 27, 40, 51, 57, 67,
        77, 82, 84, 85, 86, 87, 88, 91, 97,
        98, 102, 104, 105, 111, 118, 137,
        138, 140, 141, 144, 147, 152, 156,
        159, 161, 162, 206, 207, 208, 210,
        211, 212, 213, 217, 219, 220, 222,
        223
    fed-batch   7, 39, 40, 41
    washout   86, 87
cyanobacteria   15, 46, 134

cytoskeleton   25, 26, 29, 33, 35

2-deoxyglucose   223
DAHP   108, 109, 110
data bases   4
DBT   203
        See also Dynamic Bifurcation
            Analysis
degree of reduction 44, 45, 46, 48, 49, 51,
        67, 68, 70, 71, 73
diacetyl   8, 10, 11, 123, 124, 126
diffusion   35, 121
DNA
    microinjection   185
    -binding proteins   27
doubling time   78, 79, 112, 217
downstream processing   97, 99
Drosophila   17, 20
dynamic
    bifurcation analysis   135
        See also DBT
    instability   26, 33, 35
    organization   21, 22

E. coli   10, 11, 12, 13, 14, 16, 72, 73,
        105, 106, 108, 109, 115, 116,
        122, 138, 156, 158, 161, 167,
        180, 201, 219
eigenvalue   149
    imaginary   149
elasticity coefficient 7, 128, 130, 132, 139,
        142, 144, 201
electron acceptor 16, 42, 43, 46, 51, 71, 72,
        126, 159
elemental formula   42, 45
Embden-Meyerhoff   72, 218
endosperm   181, 188, 189, 190
energy
    uncoupling   216, 217, 218, 237
    dissipation   93, 113, 214
    spilling   69
    -limited   68, 69, 117
    potential   64
5-enolpyruvylshikimate-3-phosphate   177
enthalpy and free energy   65, 71
Entner-Doudoroff 64, 72, 85, 121, 123, 208,
        209
enzyme
    concentrations   93
    glycolytic   16, 27, 144

"regulating"   133
equation
    nonlinear differential   149, 154
    ordinary differential   36, 144, 200
erythrose-4-P   52, 53, 54, 56

fermentation
    ethanolic 13, 35, 46, 103, 105, 137, 140, 142, 210, 213, 214, 217, 218, 221
    products   119
flow cytometry   4, 99, 221
flow   22, 23, 34, 75, 80, 185
    assimilatory   80
    dissimilatory   28, 80
flux
    anabolic 31, 37, 93, 95, 111, 119, 214, 215, 216
    and metabolite concentrations   7
    catabolic 9, 14, 26, 31, 37, 51, 55, 62, 72, 80, 93,, 97, 102, 112, 118, 184, 205, 208, 214, 223
    coordination hypothesis   93, 95, 214
    redirection   93, 95, 119, 138, 216
"forced" synchrony   168
formate   70, 73, 75, 105, 124, 126
frequency-forcing   154
fructose-6-phosphate   121
functional genomics   18
futile cycles   80

G1 phase   96, 214, 217
gas chromatography/mass spectrometry   111
gene
    overexpression   7, 205
    glucose-repressible   137, 213
    SNF1   223
genome   15, 17, 18, 20, 29, 30, 184
gluconeogenesis   117, 206, 218, 223
    enzymes   20, 206, 219, 223
    substrates   19, 209
glucose consumption rate   103, 105
glucose-1-P   179
glucose-6-phosphate dehydrogenase   27
glutamate   61, 102, 104, 192, 206, 207
glyceraldehyde-3-phosphate   121
3P-glycerate   190, 191, 193
glycerol   13, 14, 20, 52, 56, 59, 70, 73, 98, 102, 103, 113, 205, 206, 209, 219

glycine   16, 57, 61, 62, 192
glycogen   63, 134, 199, 200, 223
glyoxylate cycle 110, 117,137, 205, 206, 223
glyphosate   176, 177
growth
    anaerobic 16, 32, 46, 48, 62, 67, 69, 71, 72, 119, 122, 124, 126, 159, 162, 208
    -limiting substrate 79, 80, 82, 84, 85, 86, 87, 89, 91, 157
    microaerophilic   45, 48, 208
    nitrogen-limited   102

heat of combustion   67, 68, 70, 71
herbicides   176
hexokinase   27, 133
homeodynamic   19, 21, 22, 34, 37, 204
hybridoma   16, 40, 116, 118

IME1   223
inulin   220
invertase   15, 27, 41, 218
isotopic labeling   4

kinetic
    control   100, 135
    energy   64
Klebsiella aerogenes   44, 159
Kluyveromyces marxianus   220, 222

lactate dehydrogenase   13, 124, 125, 126
lactic acid   11, 46, 123, 209
Lactococcus lactis 11, 13, 16, 123, 124, 125
leucoplasts   188
lignocellulose   121
limit cycle   19, 32, 156
linear optimization   116
lipids   15, 97, 102, 207
lysine   9, 12, 61, 115, 181, 242

macromolecular components   91, 97
malate   11, 70, 73, 193, 195, 219
malic enzyme   206
mannitol   69, 70, 178, 183
mass isotopomer analysis   111
mass transfer   82
MCA   4, 7, 40, 96, 100, 110, 127, 128, 131, 132, 134, 135, 139, 140, 141, 142, 185, 186, 196, 197, 201, 203, 209
metabolic control analysis 96, 137, 139, 156.

See also MCA
metabolic flux analysis 4, 56, 96, 111, 115,
        137, 203
        See also MFA
metabolic flux balancing   115
metabolism
        endogenous   84
        fermentative 45, 104, 137, 138, 209, 219
        respiratory 24, 45, 48, 67, 69, 72, 102,
            104, 105, 113, 162, 164, 165, 205,
            214, 218, 221, 223
        respiro-fermentative 45, 67, 102, 103,
            104, 170, 206, 207, 221, 223
methane   46, 55, 62, 70, 74
methanol   69, 70, 74, 157, 222
method
        biothermokinetic   203
        Liapounov indirect   149
        matrix   127, 133, 139, 142
MFA   4, 7, 40, 56, 57, 96, 97, 106, 111,
            113, 114, 115, 117, 118, 137,
            138, 197, 205
microtubular protein   35
mitochondrial biogenesis   216, 218
model
        cybernetic   118
        Monod   150, 152, 154, 157

NADH oxidase   16, 62
Neurospora   3
nitrosonium ions   164
nonlinear kinetics   31
nuclear magnetic resonance   4, 111
nucleotide synthesis   57
nucleotides   55, 57, 58, 97, 122, 207, 214
nutritional stress   223

objective function   116
organic acids   8, 96
organism
        chemolitotrophic   43, 46
        chemoorganotroph   42
        photosynthetic   46, 174
        phototrophic   42
organized complexity   30, 204
oscillations
        damped   19, 148, 155
        glycolytic   161
        respiratory   162, 164
        spontaneous   168, 171
        ultradian   35

oscillatory dynamics   31, 32, 33, 163
oxalacetate   63, 103
oxalate   70, 73, 75
oxalic acid   62
oxidative phosphorylation 51, 62, 72, 79, 85,
            97, 113, 215, 218
oxygen
        limitation   48, 209
        uptake   24, 105, 116, 169

P:O ratio   51
parameter optimization   139, 144
pathway
        anaplerotic pathways   117
        metabolic 3, 4, 7, 24, 31, 34, 56, 73, 93,
            97, 100, 107, 121, 123, 127, 128,
            131, 188, 192, 195, 197, 205, 209
        pentose phosphate 35, 56, 97, 117, 121,
            191
        pyruvate   8, 124
        reductive pentose phosphate 190
percolation   35
P-glucoisomerase   197
Phaeodactylum tricornutum   188
phase
        plane   36, 116, 117
        portrait   96
phenomenological coefficients   47
phenotype   17, 116, 117, 204, 209
phenotypic complexity   18
phosphoenol pyruvate carboxykinase 205
phosphoenolpyruvate carboxylase   15, 193
phosphofructokinase   11, 28, 160
phosphorylation 28, 32, 33, 57, 60, 62, 64,
            68, 84, 106, 113, 134, 180, 208,
            219, 223
phosphotransferase system   9, 10, 107, 117
photorespiration   192, 193, 195
photosynthetic 46, 174, 179, 182, 186, 188,
            189, 190, 192, 193, 194, 195,
            196, 197
phytohormones   182
plant
        C_3-   193, 195
        C_4-   193, 194
        cells 173, 174, 182, 183, 184, 185, 186,
            190, 196, 197, 202
plasmid   7, 109, 122, 126, 156, 184, 185
pleiotropic   18, 20, 205, 214, 223
polyethyleneglycol   185, 200
polygalacturonase   178

polyhydroxybutyrate 209
polysaccharides 10, 55, 91, 96, 97, 102, 119, 179, 180, 200, 207
post-translational modifications 25
potential
    phosphorylation 119
    redox 57, 106, 208
propanediol 9
Pseudomonas
    denitrificans 69, 73
    extorquens 160
    mendocina 45, 208
pulse testing 154
pyruvate decarboxylase 27, 121
pyruvate dehydrogenase 16, 27, 64, 117, 124
pyruvate kinase 11, 12, 27, 28, 36, 108
pyruvate-formate lyase 124

rate-controling steps 7, 8, 93, 134, 139, 209, 210, 213, 214
redox equivalents 92, 95
relaxation times 21, 23, 34
respiratory quotient 24, 45, 67
ribose-5-P 52, 53, 54, 56, 59
ribosomes 72, 105
ribulose-1,5-bisP carboxylase/oxygenase 190
RNA synthesis 105
Rubisco 190, 191, 192, 193, 195, 196

S. cerevisiae 7, 16, 27, 37, 56, 67, 69, 73, 102, 112, 117, 133, 136, 139, 144, 162, 167,, 170, 205, 211, 215, 220
saturation constant 80
Schizosaccharomyces pombe 217
second messengers 26, 35, 134
Selenastrum minutum 186, 187
self-organization 29, 34, 35
sensible heat 66
sensitivity amplification 35, 198
serine 57, 61, 62, 136, 192, 217
signal transduction 25, 232
*snf*1 mutant 210, 212, 213, 214, 221, 223
somatic embryogenesis 183
spatio-temporal scaling 22
specific heat 66
sporulation 37, 177, 222, 223, 225
starch 14, 134, 178, 186, 188, 196
Start 96
state properties 66

steady state 24, 31, 36, 40, 57, 65, 77, 85, 87, 91, 96, 100, 111, 114, 119, 127, 134, 138, 141, 144, 149, 151, 153, 155, 158, 160, 197, 209, 221
stoichiometric coefficients 42, 45, 47, 51
stoichiometry matrix 114
strain
    isogenic 137
    wild type 126, 137, 138, 141, 210, 212, 213, 216
Streptomyces griseus 40
substrate-inhibition 154
sucrose 12, 41, 121, 136, 178, 179, 188, 189, 190, 196, 197, 198, 222
sugar alcohols 178
sugar transport 8, 10, 121, 139, 140
sugars 55, 58, 97, 108, 110, 121, 122, 208
sulphur 41, 46, 56, 57
summation theorem 131, 132
suppresors 27

taxol 181
Taylor series 149
TDA 5, 6, 7, 96, 128, 131, 136, 137, 142
terpenoids 180
thermodynamic
    efficiency 31, 32, 33, 48, 75
    limitation 100, 101
thermodynamics 6, 30, 39, 40, 65, 74, 93, 136
    first law 39, 40, 65
    mosaic non-equilibrium 203
toluene 14, 62
topology 31, 32
transaldolase 14, 121, 122, 123
transcription 25, 26, 33, 105, 178, 206, 219, 241
transcriptional activators 27
transdisciplinary approach See TDA
transgenic
    plants 173, 177, 179, 198, 201
    potatoes 180
    tomatos 178
transient 17, 77, 119, 134, 154, 158, 159, 160
transketolase 11, 14, 108, 110, 121, 122, 123, 191, 197
translation 25, 26, 105, 178

triacylglycerol   188
tricarboxylic acid cycle   35, 97
Trichoderma   3
triose-P   190, 197
tryptophan   181

ultrasensitive   32
ultrasensitivity   199
vector   3, 114, 115, 121, 148, 184, 185
    expression   3
    -free systems   183
    -mediated   183
vinblastine   181
vitamins   39, 119, 122, 176, 181
volumetric rate   120

xenobiotics   1, 3, 6, 8, 62, 153
xylene   62

xylose   7, 9, 13, 108, 110, 121, 123, 158
xylose isomerase   121, 123
xylulokinase   121, 123
xylulose-5-phosphate   121

yield
    ATP   73
    biomass   72, 156
    growth   79, 84, 117
    heat   67
    theoretical 4, 7, 75, 79, 97, 102, 106,
        110, 112, 113, 120, 121, 122, 137,
        148, 203

Zymomonas mobilis   7, 14, 81, 121, 123